DUOETHNOGRAPHY

SERIES IN UNDERSTANDING STATISTICS

S. NATASHA BERETVAS   Series Editor

SERIES IN UNDERSTANDING MEASUREMENT

S. NATASHA BERETVAS   Series Editor

SERIES IN UNDERSTANDING QUALITATIVE RESEARCH

PATRICIA LEAVY   Series Editor

RICHARD D. SAWYER AND
JOE NORRIS

DUOETHNOGRAPHY

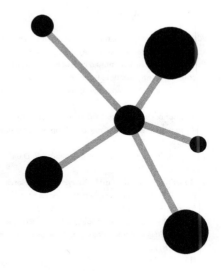

OXFORD
UNIVERSITY PRESS

Oxford University Press is a department of the University of Oxford. It furthers the University's objective of excellence in research, scholarship, and education by publishing worldwide.

Oxford New York
Auckland   Cape Town   Dar es Salaam   Hong Kong   Karachi
Kuala Lumpur   Madrid   Melbourne   Mexico City   Nairobi
New Delhi   Shanghai   Taipei   Toronto

With offices in
Argentina   Austria   Brazil   Chile   Czech Republic   France   Greece
Guatemala Hungary Italy Japan Poland Portugal Singapore
South Korea   Switzerland   Thailand   Turkey   Ukraine   Vietnam

Oxford is a registered trade mark of Oxford University Press in the UK
and certain other countries.

Published in the United States of America by
Oxford University Press
198 Madison Avenue, New York, NY 10016

Library of Congress Cataloging-in-Publication Data
Sawyer, Richard D.
    Duoethnography/Richard D. Sawyer and Joe Norris.
    p. cm.—(Understanding qualitative research)
    Includes bibliographical references and index.
    ISBN 978-0-19-975740-4 (pbk)   1. Narrative inquiry (Research method)
    2. Ethnology—Methodology.   3. Qualitative research—Methodology.
    I. Norris, Joe.   II. Title.
    H61.295.S29 2013
    001.42—dc23   2012006141

9 8 7 6 5 4 3 2 1
Printed in the United States of America on acid-free paper

CONTENTS

ACKNOWLEDGMENTS

We wish to acknowledge the support that the following have given us as we have developed this emergent dialogic research methodology.

First, we thank our colleagues and fellow duoethnographers who have helped develop and share this methodology by exploring a variety of topics in dialectical conversations with ourselves and others. The works of Sonia Aujla-Bhullar, Rebecca Bradley, Rick Breault, Deb Ceglowski, Walter S. Gershon, Kari Grain, Jim Greenlaw, Raine Hackler, Sean Hall, M. Francyne Huckaby, Donna Krammer, Deidre M. LeFevre, Tonda Liggett, Darren E. Lund, Rosemarie Mangiardi, Patrice McClellan, Morna McDermott, Maryam Nabavi, Tina Rapke, Jennifer Sader, Nancy Rankie Shelton, Kathleen Sitter, and Molly Weinburgh have provided us with many insights into the variations of duoethnography.

Second, we acknowledge a number of researchers for their support and encouragement as this methodology took shape: the Curriculum and Pedagogy Group, the International Congress of Qualitative Inquiry, the Qualitative Research special interest group of the American Educational Research Association, Mitch Allen, Jean Clandinin, John Creswell, Lisa Given, Patricia Leavy, Jan Morse, William F. Pinar, Carl Leggo, Janet Miller, and Pauline

Sameshima through their narratives, actions, and endorsements enabled this methodology to reach its current stage of development. We thank them, the editorial and production team at Oxford University Press, and the many others who indicated that they saw value in this richly emergent endeavor.

Finally we thank our partners, Nathaniel Monsour and Pauline Norris, for their continued support, patience, understanding, and feedback in all that we do.

DUOETHNOGRAPHY

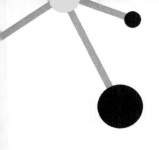

1

INTRODUCTION

AS A SOCIETY, we are approaching a condition where "there is no [human] voice to hear" (Bradbeer, 1998, p. 101). In terms of research, questions grounded in the generative and narrative core of humanity—the "naked I," the contradictory, marginalized, resistant—are usually missing in traditional projects. Left unexplored, such questions do not go away. Instead, they become both personally and socially resonant. These questions reframe research from an embodied vantage point: not from the outside in (etic)—from an external, abstract perspective—but from the inside out (emic).

For example, in their duoethnography about patriotism in the southern United States, Francyne Huckaby and Molly Weinburgh (2012), one an African American and the other a European American woman, both from the South, investigated the sociocultural context of the song "Dixie." This song, written in the mid-19th century, was widely recognized as the anthem of the Confederacy, and to this day it continues to resonate with the multiple meanings embedded in that conflict. As a still-popular song in the South, it supplied background music to Huckaby and Weinburgh's lives. In their study, they explore this song as a cultural artifact and context of analysis about perspectives on "issues of worth, dignity, power, and position"

(Huckaby & Weinburgh, 2012, p. 159). They examine relationships between individuals on the one hand and cultures, group affiliations, and institutional structures on the other. Such research foci inquire into relationships between personal values and normative rules, between restricted meanings and dominant discourses, and between subcultural and institutional practices. Such questions are embedded in the sociocultural fabric of our lives and intersect with larger principles about humanity, interrelationships, personal agency, social justice, and educational well-being. They are also purposeful, focused on empowering research participants and changing society. Research based on such questions can help humanize a culture of increasing globalization, abstraction, commodification, and standardization. The investigation of these questions calls for a methodology that mixes societal questions with individual frameworks, theory with praxis, and curriculum—that of self and school—with justice.

Part of the elusiveness of these questions, however, is that to articulate and study them requires a new way of looking at research. Duoethnography has emerged as a dialogic research methodology in response to researchers seeking to work in a new key (Aoki, 2005). A dialogic context in duoethnography is a conversation—not only between people but also between people and their perceptions of cultural artifacts (such as photos)—that generates new meanings. In duoethnography, two or more researchers work in tandem to dialogically critique and question the meanings they give to social issues and epistemological constructs. Working with a critical partner, duoethnographers select social phenomenon or themes to investigate. They then engage in cycles of interpretation which involve data analysis (often of cultural artifacts from their lives); abduction, which involves researchers' imaginative thinking about the data (Charmaz, 2009); data situation within personal stories and cultural meanings; dialogic and collaborative critique; and an articulation of new perspectives and insights. In this process, they seek not commonalities but differences as they collaboratively develop a transformative text. Examining personal and cultural artifacts (Chang, 2008), stories, memories, compositions, texts, and critical incidents, duoethnographers excavate the temporal, social, cultural, and geographical cartography of their lives, making explicit their assumptions and perspectives.

Duoethnography is loosely based on Maurice Merleau-Ponty's (1962) belief that consciousness and culture influence experience and that experiences are always mediated by individual and cultural meanings given to past experiences. Its studies are layered examinations of meanings existing between researchers' lives and their cultural contexts. Descriptive yet critical, their focus is consistent with Chang's (2008) description of the intent of autoethnography, a method of critical self-study of cultural influences on one's life: "Mere self-exposure without profound cultural analysis and interpretation leaves this writing at the level of descriptive anthropology or memoir" (p. 51). Considering themselves the sites rather than the topics of their research (Oberg, 2002), duoethnographers seek to examine and reconceptualize their narratives of interpretation—how they have come to understand an incident or theme in and through their lives as well as the ways in which they have situated (and have been situated by) this understanding temporally, socially, geographically, and culturally.

Grounded in social justice, duoethnography has been used purposefully to promote change. Some duoethnographers have sought to gain a critical awareness of their own narratives of experience through a dialogic process, thereby leading to a change in perspective or a restorying of a narrative; others to critique or deconstruct culturally dominant discourses by juxtaposing them with personal "counterpunctual" (Said, 1993) narratives; and others to complicate cultural meanings through a dialogic, collaborative lens. Duoethnographies are written in such a way that the voice of each duoethnographer is made explicit. Juxtaposing their stories, duoethnographers discover and explore the overlapping gray zones between their perspectives as intertwined intersections that create "hybrid identities" (Asher, 2007, p. 68) instead of binary opposites.

This process of collapsing dichotomies is illustrated by Diversi and Moreira (2009) in their dialogic autoethnographic study of Brazilian street children. Diversi stated:

I felt as strongly as ever that the only way to more inclusive systems of social justice was through the expansion of the dominant discourse about the other. And this discursive expansion had to be done by challenging binary systems of either/or, normal childhood/abnormal childhood, good kid/ bad kid, us/them, more human/less human, children/little criminals. Ideologies of domination, such as the one behind street children, depend on people accepting that humans can be summed up by essentializing dichotomies of self. (Diversi & Moreira, 2009, p. 15)

Working with street children in Rio de Janeiro, Diversi and Moreira began to deconstruct ways in which they themselves were implicated in projects of domination. As they examined their perceptions of themselves and the homeless children, who in many cases were drug addicts, they began to conceptualize a new sense of home and agency for those children and a new sense of corruption and disease for the state. Their study promoted a sense of praxis for them.

Foregrounding themselves as their participants in their studies, duoethnographers interpret their dialogically created meanings and seek critical tension, insights, and new perspectives. These multiple perspectives stem in part from the explicit dialogic focus of duoethnography. In duoethnography, researchers construct their narratives as they deconstruct them. In this process, they create a context for personal praxis and change as their inquiry into personal experience acts as a creation of new experience. As they deconstruct their narratives in collaboration within a dialogic process, they simultaneously reconstruct them with more complex and layered perspectives.

Dialogue within duoethnography functions as a mediating device that assists researchers in examining the frames that they use to situate meaning. Promoting heteroglossia—a multivoiced and critical tension (Bakhtin, 1981)—dialogues are not only between the researchers but also between the researcher(s) and artifacts of cultural media (e.g., photographs, songs, even the duoethnographic text itself). Much as in the reader response theory of Rosenblatt (1978), this process creates a new text through the transaction. For example, Liggett and Sawyer (2009) created a multivoiced text by juxtaposing high school yearbook photos from two periods and parts of the United States (the 1970s Northwest contrasting with the 1980s Midwest) in their duoethnography of postcolonial education. Examining images and memories from a trip she took to Africa shortly after high school, Liggett writes:

> Even in high school, while there were many clubs and organizations, there was no rallying around anything political and notions of diversity at my school meant the French and Spanish language clubs. I can only think that this provided fertile ground for shaking me to my very core as I stepped out of the Nairobi

airport and realized that it was nothing like Kansas. Little did I know that my dominant white, middle-class orientation was about to be deconstructed. (Liggett & Sawyer, 2009, p. 7)

A Conceptual Framework for Duoethnography: Social Justice, Narrative and Autoethnography, and Curriculum Theory

Duoethnography first appeared as a research methodology in 2004, when Norris and Sawyer wrote a dialogic autoethnography and selected the name "duoethnography" because of its plural foci. They began working with duoethnography to (1) (re)present their stories and (2) expose the culturally symbiotic nature of sexual orientation in a heteronormatively framed world. This particular investigation required a juxtaposition of contrasting narratives. Following this initial investigation, additional researchers began working with duoethnography. Innovative and tolerant of ambiguity, the initial duoethnographers drew from narrative research and autoethnography, inventing their method as they used it. These first duoethnographers shared a general critique of more traditional forms of qualitative research and sought to work in explicit forms of collaboration to co-construct and deconstruct their topics of investigation. Each pair (or group) of duoethnographers wanted to collaborate on the research topic and include the process of collaboration as an explicit part of the investigation itself. Furthermore, they did not want to have their prior beliefs reinforced as they engaged in their inquiry. They sought instead to create a new experiential sense of the possible.

For example, in one of the first duoethnographies, Morna McDermott and Nancy Rankie Shelton (2012) investigated their curriculum of beauty. As both a hidden and expressed curriculum, "beauty" goes to the core of North American if not Western culture. Indoctrination into cultural meanings of beauty can shape perceptions of self and others and frame motivation. Describing their investigation, they wrote:

[We are] self-reflecting and dialoging on how we developed our own curriculum of beauty...to consider the implications of the definitions that were shaped around us and by us through our life journeys. Through the duoethnographic

process our own individual histories are re-written; the exhu-
mation and re-examination of our own memories are layered
with alternative meanings through the eyes of another (each
other). (McDermott and Shelton, 2012, p. 224)

In this study, McDermott and Shelton sought to create a collaborative and dialogic study not of similarities but rather of differences
in the varied and contradictory ways in which beauty is contextually bound, institutionalized, stratified, and internalized.

The initial duoethnographers began experimenting with duoethnography as a research methodology because they were rest-
less with the confines of traditional research. Dissatisfied with the
problematic nature of representation in other forms of qualitative
research, duoethnographers have sought instead to tell their own
story, not their particular interpretation of someone else's story.
Furthermore, they have sought to contribute to the research process
and its products by investigating dialogically and dialectically, juxtaposing cultural landscapes of contrasting perspectives, and creating critical texts through the generative nature of dialogue. They
have delineated existing narratives or created counternarratives to
culturally dominant discourses; they have promoted, not denied,
subjectivity and imagination (while at the same time challenging
and disrupting them). And they have sought to unpack epistemologies embedded in their views of research topics and designs (Ellis
& Bochner, 2000; Koro-Ljungber, Yendol-Hoppey, Smith, & Hayes,
2009). As they have engaged in duoethnography, researchers have
drawn and built on the dynamic fields of social justice, curriculum
studies, and varied forms of qualitative research.

Social Justice

When framed by principles of social justice, duoethnography is both
a reflection of social justice and a method to advance it. The concept
of social justice is premised on the recognition of the urgent need to
examine power and privilege and improve societal and environmental conditions. It calls for action to remove personal, institutional,
national, and transnational structures that impoverish, disenfran-
chise, enslave, disempower, and humiliate people.

On one level, the promotion of social justice requires individuals
to critically examine both their own views and the personal biases of
others. Embodied, these views and biases are lived within personal,

cultural, and institutional narratives. The critical exposure of how one is situated in relation to narratives of injustice involves "conscientization" (Freire, 1970), a dialogue for consciousness-raising that can promote a commitment to change. Said (1993) clarifies the role of narrative in one's internalization of structures of injustice:

> Narrative is crucial... [S]tories... become the method colonized people use to assert their own identity and the existence of their own history. The main battle in imperialism is over land, of course; but when it came to who owned the land, who had the right to settle and work on it, who kept it going, who won it back, and who now plans its future—these issues were reflected, contested, and even for a time decided in narrative. As one critic has suggested, nations themselves are narrations. The power to narrate, or to block other narratives from forming and emerging, is very important to culture and imperialism, and constitutes one of the main connections between them. (pp. xii–xiii)

As Said (1993) suggested, people are socialized to internalize narratives of superiority and oppression (Marable, 2007; Sawyer & Laguardia, 2010). These processes of internalization take place daily, even in the 21st century. An important point about the successful viability of such grand narratives is that they are dynamic and constantly changing, adapting to the times, often hiding within seemingly acceptable discourses (e.g., currently the need for trade protection). Omnipresent, these images and messages are part of the cultural air we breathe. They are forces of strategic domination. Their internalization is normal, but not inescapable. A commitment to social justice calls for a critical self-analysis of this internalization of domination, colonialism, and unjust discourses. It calls on people implicated in constantly evolving patterns of personal and structural injustice to unlearn the internalization of the oppressor (Asher, 2007; Freire, 1970; Sawyer, 2010; Sawyer & Laguardia, 2010). And it calls for the institutionalization of a process of "decolonialism" in our schools to allow people the opportunity to become conscious of the internalization of oppression. As Young (1990) stated, "social justice means the elimination of institutionalized domination and oppression" (Young, 1990, p. 15).

Autoethnographers and duoethnographers cannot solve social injustice. But through their words and deeds they can advance social justice. Auto- and duoethnographers can also offer narratives of exposure and resistance to dominant discourses. Critiquing the contours of one's life as a site of sociopolitical enculturation offers a counterpunctual narrative (Said, 1993), a new way to imagine cultural worlds within the tension of standardized and often neoliberal discourses. These narratives of exposure and resistance are not the basis for new grand narratives. Rather, they complicate and unpack them (Muncey, 2010).

Narrative and Autoethnography

In the hidden garden of the de Young Museum in San Francisco there is an unusual sight. On some perfectly cut grass and surrounded by almost primordial looking palm and eucalyptus trees, a large (three story) and very open Claes Oldenburg blue and silver sculpture of a safety pin seems to float off the ground (see figure 1.1). Its 50-foot needle points at the garden café. People seem surprised when they first encounter it but then begin to reconceptualize their views of the mundane safety pin as beautiful art. And slowly, as they return to their coffee and cake, their newly complex narrative of the landscape slides back into an ordered view of the world.

Our response to the big safety pin illustrates the power of narrative to both frame and reframe how we interact with our milieu and culture. We see the pin, and its size clashes with our experience

Figure 1.1

with safety pins and maybe even our story of childhood. But we soon experience it as art and reconceptualize our perception of safety pins within a new narrative unity. This dissonant process, grounded in a personal transaction, exposes conceptualization as it transforms it. Our response illustrates one way of creating a transformative, dialogic text in duoethnography. It also suggests certain characteristics that duoethnography shares with narrative inquiry and autoethnography, both forms in which it is grounded.

Polkinghorne (1988) believes in "the importance of having research strategies that can work with the narratives people use to understand the human world." In this sense, narrative inquiry has been formed around practice, not practice around theory. Drawing from anthropology, narrative inquiry represented a stark paradigm shift from more reductionist and quantitative forms of research, seeking instead naturalistic ways of inquiring into and understanding human experience. Clandinin and Connelly (2000) offered an evolving definition of narrative inquiry:

> Narrative inquiry is a way of understanding experience. It is a collaboration between researcher and participants over time, in a place or series of places, and in social interaction with milieus. An inquirer enters this matrix in the midst and progresses in this same spirit, concluding the inquiry still in the midst of living and telling, reliving and retelling, the stories of the experiences that make up people's lives, both individual and social. Simply stated...narrative inquiry is lived and told. (p. 20)

As Behar-Horenstein and Morgan (1995) state, "Narrative helps expose the connections between what humans think, know, and do as well as the reciprocal relationship between the way that human thinking shapes behavior and knowing shapes thinking" (p. 143).

Thus narrative inquiry is shaped around how people live and make sense of their lives. It also creates narrative itself. A concept central to narrative inquiry is narrative unity (MacIntyre, 1981). Narrative unity is an impulse that generates narrative as people make sense of their lives and organize their perception of experience (Connelly and Clandinin, 1990). As people live storied lives, they organize experience into story, which allows for coherence and unity. It lets people navigate lives without, in the extreme case, the existential randomness of chaos. As narrative inquiry is lived, narrative unity underscores the organization and lived nature of such inquiry.

Narrative itself is not the outcome, but rather the origin, of change (Coles, 1989; Czarniawska, 1997). In duoethnography, narrative inquiry as both "phenomena and method" (Clandinin & Connelly, 2000, p. 18) is a key concept. Related to the notion of active narrative unity, duoethnography shares with narrative inquiry a belief in the agency of people to organize and direct their own lives, even against larger backdrops of more dominant narratives and discourses. Duoethnography, also shaped to the naturalistic contours of life, seeks to build on ways that people construct both unity and disunity.

In duoethnographic studies so far, narrative unity also has emerged as a key concept. As duoethnographers work collaboratively, they present their own narratives, formed around their inquiry themes and questions. Researchers construct narrative unities and a new sense of coherence as they engage in dialogue. These dialogues are made explicit through a script format, and as the conversations progress a growth in perspective emerges. The telling and retelling of these narratives within the emerging dialectic stimulate imagination and newer ways of perceiving the ordered coherence of the initial narratives. In self-study research, unity and disunity work together to promote one's critical and transformative reading of beliefs and understandings. It is during this critical process of deconstruction that new narratives are being formed around new unities and continuities. Thus the impulse to create narrative unity acts as a dialectic within duoethnography.

The narrative unity in duoethnography centers on the researchers, who are the site of their own inquiry, interpretations, and representations. Duoethnography shares this emphasis on a subjective focus with autoethnography (and narrative), especially in its more critical and self-reflexive forms. Stemming from anthropology (Chang, 2008), autoethnography is

an autobiographical genre of writing and research that displays multiple layers of consciousness, connecting the personal to the cultural. Back and forth autoethnographers gaze, first through an ethnographic wide-angle lens, focusing outward on social and cultural aspects of their personal experience; then, they look inward, exposing a vulnerable self that is moved by and may move through, refract, and resist cultural interpretations. (Ellis & Bochner, 2000, p. 739)

Autoethnographies may be descriptive, confessional (self-evaluative), analytical and interpretive, or imaginative (Chang, 2008). The emphasis on the individual as the site of the research is foregrounded in reflexive autoethnographies, in which "personal experience becomes important primarily in how it illuminates the culture under study" (Ellis & Bochner, 2000, p. 741). Through thick description (Geertz, 1973), autoethnographers engage in cultural analysis of the relationship between the individual and his or her sociocultural and political contexts.

Duoethnography draws from a tradition in autoethnography of researcher dialogue and dialectical development. For example, Chang (2008) states that the irony of autoethnography is that it is not about self alone but also about others. Ellis and Bochner (2000) remark that in autoethnographic "texts, concrete action, dialogue, emotion, embodiment, spirituality, and self-consciousness are featured appearing as relational and institutional stories affected by history, social structure, and culture, which themselves are dialectically revealed through action, feelings, thought, and language" (p. 739). And Roth (2005), describing an autoethnography about a science project he worked on with a colleague, noted that his awareness of their contrasting perspectives underscored the boundaries of their knowledge as culturally and historically situated.

Autoethnography as well as duoethnography focuses on "intersubjectivity[,] thereby avoiding false claims to objectivity and failure-prone inner (hyper) subjectivity" (Roth, 2005, p. 3). By critically juxtaposing their stories, duoethnographers engage in a "radical suspension of judgment and submission to a systematic method of dealing with one's own prejudices and prejudgments" (Roth, 2005, p. 9). In addition, duoethnography shares with autoethnography and narrative inquiry an intent to seek not categorical conclusions but rather exposure, transformation, and uncertainty.

Curriculum Theory

Duoethnography is also a form of curriculum and curriculum making. Drawing from Schwab's (1978) notion of the four commonplaces of curriculum, Clandinin and Connelly (1992) offer this definition of curriculum making: "A view [of curriculum] in which the teacher is seen as an integral part of the curricular process and in which teacher, learners, subject matter, and milieu are in dynamic interaction" (p. 392). The interactions of the various contexts within this dynamic support a view of emergent curriculum as illuminating, engaging, and interactive. This dynamic is also found within duoethnography.

As a curriculum of inquiry and praxis, duoethnography has developed in response to the evolution in the study of curriculum and curriculum studies. The focus of curriculum studies has shifted from an emphasis on design and implementation (e.g., the work of Tyler, 1949) to a focus on interdisciplinary and dynamic cultural texts. Building on the work in the 1930s of the social reconstructionists such as Harold Rugg (1936), the curriculum reconceptualists in the United States and Canada, led by William Pinar (1975), reframed curriculum studies in the middle 1970s. A goal of reconceptualist curriculum theory is the critique of the relationship between actual school curriculum and the broader sociocultural contexts in which schools exist. Pinar (1978) claimed that "fundamental to their [the curriculum reconceptualists'] view is that an intellectual and cultural distance from our constituency is required for the present, in order to develop a comprehensive critique and theoretical program that will be of any meaningful assistance now or later" (p. 6). By "meaningful assistance," Pinar is referring to curriculum as a collective context for individual and societal improvement. Currently, curriculum studies conceptualizes curriculum as a dynamic critical conversation dedicated to understanding "interdisciplinary configurations such as African American studies, women's and gender studies, and cultural studies" (Pinar, 2005, p. 4).

A key to understanding how duoethnography promotes transformation is found in Pinar's (1975) concept of "currere," a critical form of autobiography and curriculum studies that examines the curriculum of everyday life. We are interpreting currere as one's process of engagement within the contingent and temporal cultural webs of one's life. Critical self-examination is central to this process. In currere, this critical examination process unfolds as a regressive, progressive, analytic, and synthetic endeavor premised on the recognition that conceptualization is transtemporal and changes over time (Norris, 2008; Sawyer & Norris, 2009).

This form of self-study, then, can be regarded as a looking back to make new meanings of previous experiences and conceptions. Researchers engaged in duoethnography as well as currere examine, for example, how they have learned or acquired their beliefs or behaviors in relation to cultural constructs such as beauty, sexual orientation, gender, writing, and patterns of social interaction. As Said (1993) stated, "More important than the past itself, therefore, is its bearing upon cultural attitudes in the present" (p. 17). This explicit tracking of one's construction of beliefs can lead to a change in those beliefs and a reconception of personal narratives. Currere becomes meaningful when individuals engage in a deeply subjective and honest dialectical process. Individual positionality—the vantage point from which one critiques various temporal aspects of one's life—is critical. Echoing the process of abduction, one not only critically examines social, cultural, and political contexts of influence of the past from the location of the present but also examines the present from an imagined critique from the past. This transaction supports imaginative and generative thinking about self, others, and culture.

Framed by imagination, egalitarianism, and embodied analysis, currere gives conceptual shape and form to duoethnography. It offers a vocabulary for the analysis and exposure of personal and societal epistemologies. It provides duoethnography with a dynamic context for the generation of new questions, written in a new key to seize the imagination (Langer, 1942). And it offers the possibility of new, counterpunctual narratives that run counter to dominant and dominating discourses.

Curriculum itself is the dynamic interaction of individuals within culture. It is lived collectively and animated as people inhabit it (Bradbeer, 1998; Grumet, 1988). It springs from dialogue, imagination, spirituality, and life. Inherently, it creates a space to allow people to investigate how their personal narratives are situated in relation to other narratives, allowing for a recreation of these narratives. It gives people the collective opportunity to compose, decompose, and recompose the seemingly locked-in story lines of their cultural lives.

To be successful, this transformative process cannot be lockstep and prescriptive. A static conception of curriculum curtails its generative possibilities. Those possibilities come from curriculum's "animating vision" (Reid, 1993), from its multiple and multiply changing contexts. An animating vision of curriculum emphasizes intersubjective ways of knowing and learning, joint constructions of meanings, the exploration of metaphysical questions, and the promotion of transformative dialectics. It involves a notion of ethics and civic life within collective human action (van Manen, 1994). One level of this animating vision is found in a dynamic notion of curriculum making, like that of Clandinin and Connelly (1992), that involves manifold contexts (the classroom, school, neighborhood, cliques, structures of authority, family, religion, and so on). Another level is found in aesthetic engagements of curriculum (Becker, 1994; Greene, 2001; Leavy, 2009; Sullivan, 2005). Aesthetic curriculum , for example in school, often leads to the development of intersections of multiple interpretations, both of the nature of curriculum and within teachers and students who are involved in its formation. This transformative power of aesthetic curriculum often opens with an artistic lens that subverts one's view of the normal and taken for granted, leading to new perspectives and deeper understandings.

Generative curriculum involves a process of "mythopoesis" (Bradbeer, 1998). In a narrow sense, mythopoesis involves cultural and personal imaginal negotiations as well as their sources (Bradbeer, 1998). These imaginal sources are engaged by the combined heterogeneity of viewpoints needed to begin to "address the idea of Yin—the response of the heart to the world, our subtle receptions to our raptures and hurts; and our play with the sense of our 'dim sources'" (Bradbeer, 1998, p. 7). Our evocation of these sources provides a way of generating "strategies for taking [a] problem out of its state of unnamed disquiet" (Bradbeer, 1998, p. 30), promoting the "arts of problemation" (Schwab, 1978).

The Tenets

The central tenets of duoethnography are "living" tenets, in that they develop within use. In this discussion we place them within a three-level conceptual framework of social justice, curriculum theory, and narrative and autoethnography. In chapter 2 we discuss these tenets in greater detail in a design context by exploring examples of them in specific duoethnography studies. We offer the following tenets as working principles rather than methodological steps.

Currere as Frame for Investigation and Transformation

As can be seen from the earlier description of currere, this concept holds a couple of key ideas for duoethnography. First, duoethnography is framed as a lived curriculum, or currere, in which the dialectical process ("regressive-progressive-analytical-synthetical" [Pinar, 1994, p. 19]) is central. The goal of a dialectical interaction is not a greater understanding of existing meanings and interpretations. Rather, it is the reconceptualization of those meanings. One's active and interpersonal engagement in promoting a dialectical process is foundational to a reconceptualization of understanding. Second, duoethnography draws from currere's emphasis on learning to read self as text—as fluid, recursive, and multilayered within a cultural context. The act of recalling meaningful events and reading personal beliefs within a playful yet disciplined dialogic frame itself becomes part of the currere and subsequently the duoethnography.

Voices "Bracket In"

Duoethnography calls for articulation of the belief systems embedded in language and research as a first step in personal praxis and cultural exposure and change. As such, subjectivity and personal epistemology are a focus of study and analysis in duoethnography. In duoethnography subjectivity is not bracketed out (Spradley, 1980). Central to bracketing in is that subjective identity and personal epistemology are foregrounded as a focus of analysis: "Thus reaching an understanding is not a matter of setting aside, escaping, managing, or tracking one's own standpoint, prejudgments, biases, or prejudices. On the contrary, understanding requires the *engagement* [italics in the original] of one's biases" (Schwandt, 2000, p. 195).

Self as Research Site, Not Topic

Our narratives are inscribed and partly scripted by their historical-cultural frames. What the cultural inscriptions are, how they have been formed, what larger societal discourses and narratives they reflect, and how we negotiate them—these are all research foci of duoethnography. To explore these cultural influences and contexts, duoethnographers examine themselves as the site, not the topic, of their research (Oberg, 1992). In doing so, they examine the temporal, social, cultural, and geographical cartography of their lives. The research process is one of interrogation—not reification—of personal or socially relevant issues such as race, ethnicity, and sexuality. The duoethnographers' subjectivity, always present, creates a human frame for their "findings." Those findings are not facts; rather, they are unique and changeable presentations or representations of meaningful narratives of experience. As can be seen from the above, duoethnography scaffolds a process of critical engagement of our stories—too often our personal mythologies. Instead of viewing these intertwined intersections as binary relationships (you/me, present/past, inside/outside), duoethnography asks researchers to "to engage, instead of repress or deny, our hybrid identities, our in-between locations" (Asher, 2007, p. 68). Seeking those in-between locations, researchers are focused not on self as topic but rather self as research site in relation to lived cultural worlds.

(Re)storying Self and Other

Clandinin and Connelly (2000), referring to the work of Bateson (1994), suggested that all of us "lead storied lives on storied landscapes" (p. 8). Story provides context and an arc to one's perspective and views of life. It also provides a sense of continuity, as seen in the concept of narrative unity (Clandinin & Connelly, 2000; MacIntyre, 1981). Stories frame how we perceive the world, and our narrative unities provide a sense of continuity to those perceptions. One of the intents of duoethnography is to promote researchers' increased awareness of their stories—to shift their perspectives of their narratives from a present and personal vantage point to multiple cultural and temporal vantage points for dialogic self-examination. It is important to note that while a sense of restorying is often evident during the research and the writing of a duoethnography, it sometimes manifests at a later date. The study can create a capacity for future change and deepening of the restorying process. Researcher trust is key to this process.

Quest(ion), Not Hero/Victim

Duoethnographers try not to situate themselves as either heroes or victims. Hero and victim stances place the researcher as the topic, at the center of the inquiry process, usurping meaning making and change. Placing the self instead on an honest, critical journey toward meaning recognizes that the present understanding is incomplete. If duoethnographers enter into their research and writing with fixed ideas, the result will not be dialogic. The emphasis is on the "quest," the questioning, because the conversation with the other should change one's personal stories in some way. Hero myths resonate with subtextual discourses. Those are the meanings we seek to excavate and explore, not reify.

Fluid, Recursive, Layered Identity

The generative process of making meaning is clarified by postmodern views of self and identity. In the past two decades the theoretical study of the self has moved away from deterministic, universalist perspectives, defining individuals as autonomous, linear, bounded, transcontextual (Holland, Lachicotte, Skinner, & Cain, 1998), and binary (Said, 1993). More postmodern theories of the self portray identity as fluid, multilayered, and contradictory. "People's representations of themselves in the stream of everyday life reveal a multitude of selves that are neither bounded, stable, perduring, nor impermeable" (Holland et al., 1998, p. 29). Building on the work of Vygotsky (1978) and Bakhtin (1981), social constructivist views of self suggest that identity is formed through a mutual interplay of cultural and structural symbols and discourses as well as individual characteristics. Social cultural discourses are dynamic and in flux and "are conceived as living tools of the self— as artifacts or media that figure the self...in open-ended ways" (Holland et al., 1998, p. 28).

Understandings Not Found: Meanings Created, Exposed, and Transformed

Duoethnography is premised on the view that meanings are not fixed entities independent of the interpreter, waiting for discovery. Drawing from Schwandt's (2000) discussion of epistemological stances for qualitative inquiry, duoethnography rejects "an ontology of the real" (p. 197), "what might be called meaning realism" (p. 198). Notions of meaning realism involve an atomistic, representational account of meaning and knowledge. Duoethnography, in contrast, emphasizes "the notion of the coming into being of meaning" (p. 198). Schwandt (2000) describes this process in relation to phenomenology: "Meaning is negotiated mutually in the act of interpretation; it is not simply discovered" (p. 195). That is, meaning is not collected, but generated (Norris, 2009). Such relational research holds the possibility of "an ethical commitment in the form of respect for and fidelity to the life world" (Schwandt, 2000, p. 193). Furthermore, it promotes a process of grounded, dialogic, and participatory democracy.

Emergent, Not Prescriptive

Duoethnographies are emergent, and to avoid becoming overly prescriptive their methodologies remain open. Themes and questions emerge as each group of researchers explores meaning and story in relation to unique locations. As Cunningham (1988) pointed out, "At its simplest, there is no group process to attend to, only the interpersonal relationship of two people" (p. 164). Creativity and playfulness can help researchers both develop and sustain multiple dialogues, including those between researcher and researcher, researcher(s) and artifacts, past and present, text and text, and other unfolding combinations. Being emergent, this process requires researchers to develop specific research skills: listening critically to the research partner, interweaving inductive reason with deductive framing, and working with the co-researcher as an equal partner even in the construction of one's own narrative—to mention a few. In this process, researchers generate, expose, and revise meaning together.

Communal yet Critical Conversations as Dialogic Frame

In duoethnography, communal yet critical conversations are a central means for problematizing perspectives, dominant discourses, narratives, and narrative unities. When viewed dialogically, conversations can involve a range of dialectical interactions, such as those between people, on the one hand, and personal, cultural, historical artifacts, and temporal periods on the other hand. These conversations allow researchers to begin both to deconstruct and to reconstruct their narratives. As we examine our narratives from specific vantage points in time and space, we encounter an impulse to build and reinforce narrative unity from a self-perception grounded in a particular time and place in the face of the other (Lévinas, 1984), thus making meaning in the present filters of our perceptions of the past. Critical conversations provide structured heteroglossia (Bakhtin, 1981; Holquist, 1981)—multivoiced and critically tense sites—to promote both disruption and reformation of these discourses, narratives, and perspectives.

Trust and Recognition of Power Differentials

The creation of a safe space for researchers is central to duoethnography. While traditional ethnographers attempt to uncover their participants' meanings, duoethnographers attempt to interrogate their own meanings. Researchers involved in duoethnography can expect that their and their partners' stories will generate counternarratives (Neilsen, 1999). With this can come personal risk. In some cases, researchers begin their study from a point of long-standing trust and collegiality. Trust respect, and collegial relations, however, become more complicated when power differentials found in society are present between the researchers. The inequitable conditions in the "real world" can be (re)inscribed by researchers as they conduct a duoethnography. If researchers have not previously worked together and power inequities exist between them, they can, as a starting point in their work, begin building trust by establishing "rules of the road." These rules can cover everything from researchers' roles as they collect data to their participation in the analysis and reporting of their emerging narratives.

Place as Participant

Place is considered an active participant in duoethnography. Place is often thought of as physical location. In duoethnography we attempt to understand and experience our sense of place as something much more complex. Duoethnography recognizes place as contributing to interactions, conversations, and transactions—leading to mutual change (in both place and ourselves). We seek to understand place as evolving layered narratives of multiple cultures, landscapes, and place identities (dynamic, depressed, university town, family friendly, etc.). Each place has a unique focus and to an extent frames inhabitants' identities. For example, if you've had the luxury to live in different places, consider how the places changed you. Each location has its own evolving place narratives, which you become a part of and contribute to. Furthermore, each place has its own epistemologies and ways of knowing, and it is these epistemologies, if we become aware of them, that contribute to complex and multifaceted ways of knowing.

In duoethnography, we tell stories that examine our relationships to evolving narratives (again, the holistic combination of places and their changing culture[s], landscapes, history, and identity). As we do this, we attempt to increase the complexity with which we "read" a sense of place. This process might involve a critique of personal and cultural epistemologies in relation to place. As we contemplate our relationship to a particular place, we consider the relationships we have had with other places and attempt to construct a personal frame that involves a complex ecology of interconnected places. And as we construct a new way of interacting with and learning from place—a pedagogy of place (Greenwood, 2009)—we seek to construct a new ethic of place.

Literature as Participant

As in other forms of qualitative research, literature reviews in duoethnography serve multiple functions, such as offering a conceptual framework and presenting a theoretical context for critique and discussion. Typically, though, and in contrast to other qualitative forms of research, the literature review in duoethnography is blended into the written study, and literature is often critiqued as a cultural artifact. The literature review is not conducted before the onset of a study. Instead, as the topic emerges, researchers investigate and introduce pertinent and meaningful literature.

Greenlaw and Norris (2012), for example, offer this example of a blended literature review in their duoethnographic study on the curriculum of writing:

> Our dialogue about our emergence as writers differs from, say, the collection of letters written between Fitzgerald and Hemingway (Donaldson, 1999). Like us, they were friends, but they were also in competition with each other. Their epistolary dialogue, which took place over several years, was

written for the benefit of themselves as writers. In contrast to their dialogue, our conversation is intended to reveal the insights that we have each gained about the inspirations for our writing processes, so that we can share these revelations with our fellow educators. (p. 92)

As elucidated in this quote, the literature both frames their study and provides a contrast for its analysis.

Other duoethnographers have been more traditional in their use of literature to frame their studies. For example, Krammer and Mangiardi (2012), in their duoethnography on the hidden curriculum of schooling, first presented a detailed discussion about theories related to the hidden curriculum, narrative, and currere and then compared and contrasted their study with those concepts. They concluded their study by contributing their original concept of the "cryptic curriculum" to the professional literature on the hidden curriculum.

Difference as Heurism

In duoethnography, contrast and difference are viewed as strengths, not deficits. Duoethnography does not seek universals. Instead, it examines how different individuals give both similar and different meanings to a shared phenomenon. In this process it looks to the margins to create a range of meanings (Kincheloe, 2001). The gaps that lie between the different perspectives and different voices create a space for the exploration of the fluidity of identity, the layered, recursive, often inconsistent layers of who we are. Instead of finding the mythology of our lives, we seek to unpack the hybrid nature of our identity. This multiplicity may be found in the "in-between" spaces and the differences between us.

Reader as Coparticipant and Active Witness

The script format enables the reader to perceive the stories in the researcher and the researcher in the stories. The authentic voice of the writing is invitational, asking the reader to judge and respond with a new text. This new text is a transaction among the reader, the writer, and the written or spoken study (Barone, 1990; Rosenblatt, 1978). As Roth (2005) stated, referring to Derrida

(1985), "biography (ethnography)...comes to life only through the countersignature of the reader or hearer" (p. 11). This countersignature can offer critique and a moral compass to the study:

> The usefulness of these stories [autoethnographies] is their capacity to inspire conversation from the point of view of the readers, who enter from the perspective of their own lives. The narrative rises or falls on its capacity to provoke readers to broaden their horizons, enter empathically into worlds of experience different from their own, and actively engage in dialogue regarding the social and moral implications of the different perspectives and standpoints encountered. Invited to take the story in and use it for themselves, readers become co-performers, examining themselves through the evocative power of the narrative text. (Ellis & Bochner, 2000, p. 748)

As readers are aware of this process, so are the writers. Similar to Brecht's (1957) alienation effect, as writers engage in critical storytelling, not only exposing personally meaningful details but engaging in a societal critique, they are watching themselves and cannot hide behind "obfuscating claims of objectivity" (Lather, 1986, p. 66). The triangular relationship between the reader, the writer, and the text—playing out publicly often in presentations or readings of written studies—creates a greater level of critical dialogue within the study, promoting exposure, praxis, and a more critical restoration of researcher narrative. It asks the reader to create "catalytic validity" (Lather, 1986), a form of validity grounded in readers' resonance to a study and their degree of intended application of its meanings. As Patti Lather (1986) states, "the research process re-orients, focuses, and energizes participants in what Freire (1973) terms 'conscientization,' knowing reality in order to better transform it" (p. 67). For the writer, this process provides a witness to the integrity of the research process and the ethics of the purpose of the study.

Toward a More Holistic and Inclusive Vision of Ethics in Research

The definition of ethics in Western philosophy "holds that the virtues (such as justice, charity, and generosity) are dispositions to act in ways that benefit both the person possessing them and

Table 1.1
Summary of Duoethnography Tenets

Tenet	Summary
Currere as frame for investigation and transformation	Currere is a concept that views life history as an informal curriculum. It promotes individual and collective meaning making through personal analysis and reconceptualization of thought and action.
Voices "bracket in"	Inquirers position themselves in the text, not outside it, in neutral and objective ways.
Self as research site, not topic	The self is not the focus of inquiry. Rather, it offers a context for the analysis of socio-cultural factors affecting one's experience.
(Re)storying self and other	A goal in duoethnography is inquirers' reconceptualization of their narratives of experiences that have been important in their lives.
Quest(ion), not hero/victim	Duoethnographers seek to explore questions of importance in their lives but not to place themselves in roles that limit and close thought, such as hero or victim.
Fluid, recursive, layered identity	Duoethnography is premised on postmodern notions of identity that view self as layered, nonlinear, contradictory, emergent, and always open to uncertainty and change.
Understandings not found: Meanings created, exposed, and transformed	Duoethnographers explore and create meanings in a process of dialogue. They do not find meanings as objective truths or realities.
Emergent, not prescriptive	The goals of inquiry are not predefined outside the project and applied to it. Its ends and means emerge together.
Communal yet critical conversations as dialogic frame	Duoethnographers have an obligation to promote each other's contrasting—not similar—views about themes under consideration.

(*continued*)

Table 1.1 (*Continued*)

Tenet	Summary
Trust and recognition of power differentials	Societal power differentials are taken into account and addressed explicitly if necessary.
Place as participant	Place-based meanings—such as geographic narratives, stories, politics, and evolving culture in specific local and global spaces—are considered part of the inquiry.
Literature as participant	Theory and professional literature inform the inquiry but also present a context of analysis within the study.
Difference as heurism	Working in tandem, duoethnographers juxtapose narratives of difference, not similarity, to examine and open new perspectives on experience.
Reader as coparticipant and active witness	The readers of a written duoethnography and the audience members of a presented one are recognized as active coparticipants and meaning makers in the emergent process.

that person's society... [, that] the concept of duty [is] central to morality... [, and that] the guiding principle of conduct should be the greatest happiness or benefit of the greatest number" (*The Oxford Online Dictionary*, 2010). While these elements of ethics may seem clear, they are actually open to a range of interpretations when applied to research. The range of approaches to ethics in research reflects the different definitions and purposes found within various forms of research. Researchers' perceptions of ethics in their areas are framed by how they consider the purposes of their work and their epistemological stances toward it. Modernist and positivist definitions of research are premised on an external view of reality in which research content can be objective and neutral and therefore value free. Rooted in changing views of science from the Enlightenment, a belief in an external location of factual knowledge allows researchers to situate their work as autonomous from sociocultural contexts. Recent applications of this perspective have been influenced by the work of Max Weber (1949), as noted by Christians (2005): "Given Weber's importance methodologically and theoretically, for sociology and economics, his distinction between political judgments and scientific neutrality is of canonical status" (p. 142). The claim of scientific neutrality allows researchers, often unknowingly, to view their work as independent from and not responsible for social conditions. This claim allows an ethical goal of research to be its contribution to scientific knowledge, independent of the possible uses of that knowledge.

Research ethics within an objectivist epistemological view of science revolve around notifying autonomous research subjects of the meaning of the research project and requesting their consent to specific behavior and actions. Research organizations (e.g., the American Psychology Association) and institutional review boards for universities have established ethical expectations in research consistent with such a view of science and research. These principles include informed consent of research participants, a lack of deception, a guarantee of privacy and confidentiality, and accuracy in data findings and reporting (Christians, 2005). However, a number of tensions and dilemmas have arisen within an objectivist framing of these principles. For example, individuals from subjugated groups rarely research those from dominant groups. Christians (2005) notes that because participants are often excluded from research design and purpose decisions, a possibility of deception by omission exists. Also, guaranteeing participant privacy is nearly impossible, and accuracy is ideologically defined (and not recognized as such): "In an instrumentalist, value-neutral social science, the definitions entailed by the procedures themselves establish the ends by which they are evaluated as moral" (Christians, 2005, p. 145).

In contrast, critical, poststructural, and self-reflexive forms of qualitative research seek to bring ethics and epistemology together (Christians, 2005). More than 25 years ago, Lather (1986) called for researchers "to formulate approaches to empirical research which advance emancipatory theory-building through the development of interactive and action-inspiring research designs" (p. 64). Nested in social responsibility, the intent is for research to create a just society that benefits everyone, especially the most vulnerable of any particular group. As Denzin (2003) encouraged, qualitative research must acknowledge its pedagogical and political nature. This intention is nested in Freire's (1970) concept of empowerment, in which the goal is not to take power away from some but then give it to others while maintaining a corrupt and corrupting system; rather, the aim is "participation of the oppressed in directing cultural formation. If an important social issue needs resolution the most vulnerable will have to lead the way" (Christians, 2005, p.156). Freire (1970) recognized that, "Only power that springs from the weakness of the oppressed will be sufficiently strong to free both" (p. 28) the oppressed and the oppressor but warned that "the oppressed must not, in seeking to regain their humanity (which is a way to create it), become in turn the oppressors of the oppressors" (p. 28).

How researchers situate themselves in relation to their project is also value laden. Do they place themselves outside the context of their study, or have they included themselves within it? Is their self-reflexivity a research possibility? While abstraction and atomistic isolation of "variables" in research may lead to controlled findings, intersectionality, in contrast, may lead to uncontrolled and unpredictable outcomes, such as personal praxis and very particular framings of narrative understandings. Christians (2005), referring to the work of Denzin (1997), discusses a feminist communitarian approach to research as "an antidote to individualist utilitarianism" (p. 150). In this view, an ethical commitment to research and praxis is found by researchers' specific engagement in nurturing their and other people's human identity in "a mosaic of particular communities, a pluralism of ethnic identities and worldviews intersecting to form a social bond" (p. 151).

Many researchers have begun to work in duoethnography as a way to resolve or at least lessen these ethical dilemmas. In contrast to standing outside a sociocultural context, they stand within it and are accountable to it by engaging in a dialogic form of research. Their "findings" come from a research process that builds community as it models new ways of perceiving narrative scripts within the world. Instead of researching the other, researchers' dialogic self-investigation of cultural influences on their lives leads to a process of self-reflexivity, of describing and living the interrogation of their own assumptions and beliefs as they conduct their research (Mills, 2007).

Often a sense of morality is missing from definitions of ethical conduct in research. How do researchers "honor a moral obligation to cultivate their humanity" (Hellbeck, 2010, p. 26)? Researchers working with autoethnography, narrative inquiry, and duoethnography who situate themselves in their research and their research in issues and contexts of social justice often attempt to promote a sense of humanity and morality within their work. Duoethnographers who engage in dialogue with other researchers work in contingent, relational, and deeply meaningful situations.

Conclusion

Duoethnography is grounded in the commitment that social justice theory, curriculum theory, and self-reflexive and dialogic forms of qualitative research evince for necessary action leading to social change. Those of us who have been raised in countries with histories perpetuating colonialism and injustice struggle to give voice to meaningful stories of authenticity, resistance, and justice. We face historical, structural, institutional, personal, and political obstacles at every turn as we work against repression. As we seek new ways of thinking about and enacting justice, we face global organizations that destroy ancient ways of knowing and doing. Duoethnography presents a framework that researchers can use to work together to explore and change embodied epistemologies favoring conquest, consumerism, and the attempted justification of the perpetuation of poverty. Telling and critiquing stories around critical issues and questions is a form of research whose "findings" are filtered through the human voice and illuminated by one particular moment in time.

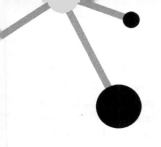

RESEARCH DESIGN

FRAMING DIALOGUE, UNCERTAINTY, AND CHANGE

LIKE ALL RESEARCH, duoethnography has an intentional design (i.e., a researcher stance that guides the work); this intentionality, however, is less about technique and more about dispositions to research. Informed by ethics, currere, and dialogic collaboration, duoethnographers consider how to make their studies trustworthy, how to use themselves as sites of study and interrogation, how to construct dialogic transactions that illuminate and problematize their topics and their thinking, how to generate deep and meaningful critiques of self and society, and how to present these in ways that invite the reader into the conversation.

On another level, intentionality in duoethnography manifests itself in the researchers' hearts and hands. It is about how they construct a study that lives in the world and promotes a deeper sense of humanity. Duoethnographers locate the purpose of research in its responsibility to society and to the improvement of existence and of place not only for people but for all living creatures and the environment. Toward this goal, they consider how their work is political and is embedded and implicated in forces of individual and institutional power and dominance. They consider how they intend to use their work as researchers, and they encourage others to use it. And they consider how their studies contribute to light in dark times (Ayers & Miller, 1998). This vision is underscored by the work and words of Maxine Greene (1991), which inspire compassion and a sense of humanity as they call all people to conscious action: "Human beings who lack an awareness of their own personal reality (which is futuring, questing) cannot exist in a 'we-relation' with other human beings. They cannot know what it means to live through a vivid present in common with another, to share another's flux of experience in inner time" (p. 8). In duoethnography, quality in design is found in its ethical stance as well as in its method.

Design Considerations in Qualitative Research and Duoethnography

As a form of qualitative research, duoethnography incorporates many of its paradigmatic design qualities. Describing these time-honored qualities, Wiersma (1991) states:

> Qualitative researchers, for the most part, do research in natural settings; they do not manipulate or intervene (except possibly by their presence) in the situation. Therefore, research design requires flexibility and a tolerance for adjustment as the research progresses. Smith and Glass (1987, p. 259) refer to this as a *working design*, similar to what McMillan and Schumacher (1989, p. 179) call an *emergent design*. From the identification of the research problem, decisions must be made about beginning the study. (p. 82)

Similar to other qualitative research designs, duoethnographies include the researchers' tentative identification of an initial problem and their flexibility and tolerance for adjustment to an emergent design. Consistent with Janesick's (2000) focus on holistic meaning found in multiple cultural contexts, qualitative research examines identity and personal relationships within systems or cultures, focuses on understanding (but not in order to make predictions), demands that the researcher stay in the setting(s) over time, casts the researcher as instrument, requires ongoing data analysis, and involves the construction of authentic and compelling narratives of understanding. And while duoethnographers reflect on naturalistic settings, they also intend to intervene and act in those settings based on their reflections.

Design in duoethnography is grounded conceptually in key constructs—some of which differentiate it in degree from other forms of
qualitative research. These include ethics, research purposes, collaboration, exploration or investigation of research questions and themes,
criteria of research trustworthiness and validity, and self-reflexivity.
Additional emergent design elements contribute to an unfolding and
uncertain process, similar to Gadamer's (1975) hermeneutic circle
of understanding. In duoethnography, however, this process is not
a closed circle. Rather, it unfolds as a constantly shifting and het-
eroglossic web of language and meaning, triggering researchers to
reconceptualize narratives and perceptions. This process is recursive
and is framed by Pinar's (1975) notion of currere as an examination
and interrogation of dialectical relationships in one's life.

Framing Design Concepts

As an emergent research method, duoethnography does not have
prescriptive design criteria, but it does have consistent framing constructs and more flexible design components. It is helpful to consider these as two levels of design. The key framing concepts exist
consistently in all duoethnographies. The flexible design components vary according to research context, problem identification,
and researcher collaboration. In this section we discuss the framing
concepts, followed in a subsequent section by the design components (Table 2.1).

Table 2.1
Framing Design Concepts

Concept	Summary
Ethics	Like all research methods, duoethnography is value laden and situational. As such, relational and situational ethics frame and guide it both conceptually and procedurally.
Research questions	Often initially uncertain, relationally contingent, and deeply embedded in experience, research questions emerge within and illuminate the inquiry.
Words of trust	Trustworthiness, an aspect of inquirers' experiential engagement in the process, is found within their reconceptualization and transformation of thought and action.

Ethics

Ethics in duoethnography are grounded in how and why researchers work together, how they consider co-participants in their study, and how they situate their study in relation to the real world. Given duoethnography's relational form, which is focused on authentic transformative collaboration, relational ethics are critical to its successful and meaningful design. The goal is for researchers to work together in ways that promote honesty and trust. Honesty precludes truth claims, manipulative arguments, and cover stories. Relational ethics in duoethnography are found in researchers' commitment to each other. This commitment exists when researchers listen actively and responsively to each other, respect each other's complex identities and cultural understandings, and recognize the limits of their study in relation to their research partner's life.

Since duoethnographers are the writers, they can control the degree of personal disclosure in their work. This process involves an awareness of power differentials, especially when power inequities within the study mirror those found in society. Relationship inequities may be lessened between researchers who have a pre-existing relationship and are used to discussing contentious or controversial issues. Researchers who are working together for the first time are often more explicit in how they approach equity within their work, in some cases even establishing "rules of the road" for their work together. In these situations it is sometimes necessary for researchers to discuss the collaborative context of their mutual work, both initially and in progress throughout their study. Darren Lund, for example, created a duoethnography with Maryam Nabavi (Lund & Nabavi, 2008), who was a graduate student at his university at the time of their study. One problem they investigated was the impact of power differentials in duoethnography. As they first discussed the nature of their work together, they considered and then constructed structures to ensure ethical collaboration.

Instead of conducting research on the other—turning someone else into an object of study—duoethnographers investigate themselves and ground their work in relational ethics. Thus in duoethnography the researcher is also the subject. In terms of ethics, this change from subject to researcher promotes participant inclusion and voice in the study. It also creates the heart of the study. When duoethnographers refer to participants in their stories, they take the ethical stance of telling a story as their own point of view, recognizing that there are different perspectives on the story. However, there are public and private boundaries that are problematic, and it is recommended that duoethnographers take extreme care when presenting or representing others.

In addition to relational ethics, situational ethics are critical in duoethnography. As duoethnography is rooted in authentic sociocultural contexts, its purpose is not separated from experience or life: It is a part of it. Design in duoethnography recognizes that its value and meaning are found in its contribution to the improvement of life experience. For example, one of us conducted a duoethnography with Tonda Liggett (Sawyer & Liggett, 2012) exploring colonial foundations of our curriculum as teachers in the United States. We examined artifacts from our lives—photos from elementary school, high school yearbooks, and our own lesson plans as K–12 teachers—in an attempt to discover colonial patterns of interaction that inscribed our lives and permeated our work. Our purpose was to identify those patterns in order to engage in more self-reflexive and ethical teaching. We found that as an interactive, intersubjective experience, duoethnography carried the same moral obligations as daily life. In our study we did not represent our lives; rather, through the presentation of our dialogic engagement, we tried to show ways of becoming more conscious of how our curriculum reflected historical patterns of colonial domination. In this way, the duoethnographic process promoted decolonialism (and postcolonialism), not colonialism. The meaning of the study lay in the unfolding process of conscientization (Freire, 1970).

In another study, Morna McDermott and Nancy Shelton (2008), drew from the work of William Doll (1993) and articulated how the duoethnography framework promoted a "recursive retrospection" and a greater sense of research trustworthiness as they worked together as duoethnographers. The relational aspect of the framework deepened the dialogic research process that lay at the heart of their duoethnography. They wrote in their study: "Duoethnography is the process of knowing the OTHER person in relationship to self as a process of research.... That's cool" (p. 11).

Research Questions

As with other forms of research, duoethnographers use questions to drive their inquiries. In duoethnography the research questions are often initially liminal and ill defined; they emerge from the stories, coming to frame them. There is always a sense of "the quest" within the question formation and exploration process.

Words of Trust

As Creswell and Miller (2000) note, "Qualitative inquirers need to demonstrate that their studies are credible" (p. 124). For duoethnography to have value for its readers and not just its researchers, readers have to believe that the researchers are not just "telling stories" (Iannacci, 2007). The issue of validity has perhaps grown more pronounced as qualitative research has become more critical, collaborative, contextualized, self-reflective, and uncertain (Creswell & Miller, 2000). The complexity of ensuring validity in these newer forms of research partly stems from problems with language. The modernist and postpositivist language of validity, stressing "rigorous methods and systematic forms of inquiry" (Creswell & Miller, 2000, p. 125), forecloses or at least limits inquiry employing more critical forms of qualitative research.

However, as qualitative research has become more critical, a search for language to reframe and position research in a new key has developed around these postpositivist perspectives. Two interrelated issues emerged as key in qualitative research: The first was how research itself could act as a nexus for praxis, researcher self-reflexivity, and social justice. The second was the "crisis of representation."

As an example of the first issue, in the mid-1980s Patti Lather (1986) discussed how language could be used to make praxis and social change the "measure" of value in qualitative research. She mentioned a new form of validity, catalytic validity, that would establish the validity of research not purely through internal research measures (e.g., rigorous and systematic controls) but rather through its purposes and goals. The value would be in the quality of research to allow participants to know "reality in order to better transform it" (Lather, 1986, p. 67). "My argument is premised," she stated, "not only on a recognition of the reality-altering impact of the research process itself, but also on the need to consciously channel this impact so that respondents gain self-understanding and, ideally, self-determination through research participation" (p. 67). Whereas Kurt Lewin (1948) had already in the 1940s called for the research process to be a nexus for societal improvement, Lather and other researchers were calling for research to be a location for self-reflection and societal change.

In the mid-1980s, the crisis of representation in research, the second of these key issues, "necessitate[d] a radical transformation in the goals of [researchers'] work—from description to communication" (Ellis & Bochner, 2000, p. 748). This crisis coincided with social constructivist views that acknowledge that people construct knowledge through their active and mutual interaction in cultural contexts (Kurosawa, 1950). Since that time, notions of validity have continued to expand to reflect research that is becoming more critical and subjective. Around these issues a new language of credibility and trustworthiness has formed to support critical, collaborative, and situated inquiry into the contradictory and dialogic nature of experience and identity.

These two issues—research as a site both for self-reflexivity and for authentic (re)presentation—became core in demonstrating the validity and trustworthiness of qualitative research. They may be found at the intersections of a number of critical theoretical perspectives (e.g., feminist theory, postcolonial theory, queer theory, and cultural studies) with forms of qualitative research as new sites for critically deconstructing and changing dominant societal discourses and narratives. More directly related to duoethnography, they are found in critical narrative research (Iannacci, 2007) and critical autoethnography (Chang, 2008).

Creswell and Miller (2000) mentioned the array of validity perspectives and terms in qualitative inquiry: "Multiple perspectives about [validity] flood the pages of books...and articles and chapters.... In these texts, readers are treated to a confusing array of terms for validity, including authenticity, goodness, verisimilitude, adequacy, trustworthiness, plausibility, validity, validation, and credibility" (p. 124). The confusion partly stems from the implicit nature of researchers' perspectives and epistemological stances in their work. As mentioned in chapter 1, in duoethnography researchers make their overall epistemologies and underlying assumptions explicit throughout their research, from their initial planning to their reporting (Koro-Ljungber, Yendol-Hoppey, Smith, & Hayes, 2009). This clarifying process helps researchers by guiding them in a process of conceptual coherence. By conceptual coherence we do not mean an a priori certainty about the project but rather an understanding about how the conceptual pieces of a project work together holistically, often in unpredictable ways. This process in duoethnography also gives the reader access to the researchers' values, beliefs, and epistemological assumptions, providing the reader with a transparent basis for making decisions about trustworthiness.

Creswell and Miller (2000) suggested a two-dimensional framework that can provide researchers with a rationale for their choice of procedure "based on who assesses the credibility of a study and their own philosophical positions toward qualitative inquiry" (p. 124). Thus the researchers make their choices explicit to allow participants, reviewers, and readers of published research to assess the inquiry based on their own epistemological stances toward it. One dimension is researcher lens, or the viewpoint "the inquirer uses...for establishing validity in a study" (p. 125). The second dimension is the researchers' paradigm assumptions or worldviews (Guba & Lincoln, 1994, as cited in Creswell & Miller, 2000) that frame their selection of procedures. Creswell and Miller list three paradigms in qualitative research: postpositivist, constructivist, and critical. Characteristics inherent in the postpositivist (or systematic) paradigm include providing data triangulation to allow readers to expand and check the subjectivity of the researcher lens, member-checking to promote validity and trustworthiness, and an audit trail to give readers or evaluators a basis for external evaluation. Characteristics inherent in the constructivist paradigm include providing disconfirming evidence to allow readers to expand and check the subjectivity of the researcher lens, prolonged engagement in the field to promote validity and trustworthiness, and thick description to give readers or evaluators a basis for external evaluation. And characteristics inherent in the critical paradigm include promoting researcher reflexivity not as a check for subjectivity but rather as a recognition of its meaning, collaboration to promote trustworthiness, and peer debriefing as a basis for evaluation. We view duoethnography as falling within and extending the critical paradigm. Duoethnography promotes researcher reflexivity through a dialogic, intersubjective process; trustworthiness through reflexivity (i.e., shared voice, co-deconstruction, reconstruction, and reconceptualization); and evaluation by providing thick description, transparency, reviewer participation, and generativity.

Because duoethnography emphasizes subjectivity, self-reflexivity is central to validity or trustworthiness. This view of reflexivity differs from Creswell and Miller's (2000) discussion of it. In their view, reflexive researchers

> self-disclose their assumptions, beliefs, and biases. This is the process whereby researchers report on personal beliefs, values, and biases that may shape their inquiry. It is particularly important for researchers to acknowledge and describe their entering beliefs and biases early in the research process to allow readers to understand their positions, and then to bracket or suspend those researcher biases as the study proceeds. This validity procedure uses the lens of the researcher but is clearly positioned within the critical paradigm where individuals reflect on the social, cultural, and historical forces that shape their interpretation. (p. 127)

As a first step in demonstrating trustworthiness in duoethnography, researchers do, consistent with Creswell and Miller, "acknowledge and describe their entering beliefs and biases early in the research process." But they continue to do so throughout the process. And in contrast to other forms of qualitative research, in duoethnography researchers bracket themselves *into* all aspects of their research text, not out of it. It is by the researcher entering the research that self-reflexivity can become a demonstration of trustworthiness. The research becomes believable when the researcher shows—through transparency and thick description—an embodied engagement in praxis and conceptual change. It is the demonstration of duoethnographers' reconceptualizations that makes the inquiry authentic and believable. The process, then, affects the concept of triangulation in duoethnography. In general, triangulation in qualitative research is a method to provide multiple forms of evidence to support central findings and themes. Duoethnographers, in contrast, are concerned more with the margins of experience (which can destabilize central meanings) than with the center. In duoethnography, triangulation as a concept implies that researchers *show* multiple forms of critical praxis and reconceptualization (or their future possibility).

An illustration of veracity and self-reflexivity may be found in Patrice McClellan and Jennifer Sader's (2012) study on interrogating power and privilege in leadership education. They begin by stating their intention to

> interrogate how discussing privilege in leadership education affects our pedagogical approaches, influences our interaction with learners, and forces us to confront our own biases when discussing power, privilege, and equity. To gain a deeper understanding of classroom dynamics around power and privilege, we, a Black woman (Patrice) and a White woman (Jennifer), revisit our lives...to uncover invisible barriers (such as fear of betrayal, feeling protective of family and friends, concerns about the proper role of self-disclosure, and students' resistance) that impact honest discussion of diversity and social justice issues in leadership education.... [In this study we] use an alternative and progressive stance on advancing the teaching of leadership and being inclusive to diverse perspectives. As leadership educators, we challenge ourselves to engage rather than deny or repress, differences that emerge at the dynamic intersections (Asher, 2007) of power, privilege, difference, and/or equity. (pp. 138–139)

This quote shows that McClellan and Sader have positioned themselves into their inquiry as a means of engaging rather than denying difference. They critique their own narratives and examine this topic dialogically—presenting contrasting narratives and expanding each other's perspectives on the topic from the periphery. In their discussion they alternate describing childhood narratives of enculturation (vis-à-vis their friends' and families' experiences with race) with present-day structural narratives (e.g., Black faculty members as an automatic faculty default for teaching diversity courses). They then apply some of their insights to a hypothetical situation: the challenges of creating a safe classroom for their students. As they discuss that topic, they make transparent their thinking. Research trustworthiness and believability are found in their embodied exploration of the tensions, which they create, among their contrasting narratives and discourses. Their research verisimilitude is embedded in their authentic, exposed voices. It is found in the uncertainty of their "findings," resonant of the uncertainty of identity.

Throughout their paper, McClellan and Sader work to make their thinking explicit and transparent. In another duoethnography, Rick Breault, Raine Hackler, and Rebecca Bradley (2012) more explicitly examine the question of transparency (and rigor) in their investigation of male elementary teachers. One of their two stated research questions asks about research validity, "How do we more transparently represent the insights gained as a result of a complex interplay of dialogue, introspection, reflection and consideration of the research literature?" (p. 121). Second, they wonder if the written report of their investigation, which sorts their conversation into themes, tends to obscure the transparency found in the transcripts of their conversations, possibly reducing the appearance of rigor in the study. Their concern was whether their written, edited document obfuscated their organic, unrevised decision-making process. Responding to these questions, they decided to use technology in a new way to promote the communication of transparency in their work:

> To accomplish our goal of creating a naturalistic and transparent presentation we decided to use the hypertext capabilities of web-based publication. All the work described below can currently be found at the Trioethnography link on http://breaultresearch.info. The materials included at that site are crucial to understanding our conversation and what we gained from it. (p. 126)

The validity found in the work of McClellan and Sader (2012) and Breault, Hackler, and Bradley (2012) coincides with Saukko's (2005) notion of contextual, dialogic, and self-reflexive validity. Here, according to Nabavi (Nabavi & Lund, 2010), researchers build validity as they "demonstrate the ways actions are informed by historical, political and social contexts,... explore the dialogical space between [themselves] and subject,... [and] analyze the ways that social processes mediate [their] experiences and environment" (p. 13).

Flexible Design Components

As we have mentioned, design considerations in duoethnography reflect its emergent research nature. As Varela, Thompson, and Rosch (1991) stated, "It [research design] is laying down a path and walking" (p. 237). In duoethnography, design and method are more intertwined than they are in other forms of qualitative research. The way researchers create and situate themselves within a dialogic research text as data distinguishes duoethnography (and its design) from most other forms of qualitative research. As duoethnographies unfold as intersubjective texts, the researchers become both the research instrument and the focus of the investigation. Collaboration itself as a form of research demands a particular dispositional stance (e.g., intense listening skills, capacity for acumen, a serious consideration of the views of others, flexibility, and imagination). Before beginning a study, duoethnographers think about their (a) constructions of multiple dialogues, (b) ways of engaging multiple forms of data, (c) ways to surface deep meanings, (d) ways to collaborate, (e) diverse conversational approaches, and (f) views of readers or audience members as participants in their work. In design, the goal is for researchers to create relational texts that promote disciplined yet free-flowing and cyclical dialogic processes that surface and disrupt narrative unities. Through their relationship, duoethnographers reconceptualize and restory their narratives of experience. As they engage in these relational processes, they use—in differing ways—a number of flexible design components. These include using multiple dialogues, surfacing subjugated knowledge, engaging in critical collaboration, finding synergy between data collection and analysis, and inviting the countersignature of the reader (Table 2.2).

Multiple Dialogues

Duoethnographies emerge in an uncertain, interconnected "rhizomatic meshingness" (Wiebe et al., 2007). This meshingness derives from the multiple dialogues and emergent ways of working within duoethnography. Seeking neither polarities nor commonalities, duoethnographers create dialogic tension as they construct, deconstruct, and reconceptualize their narratives of experience in tandem with a research partner who is also engaged

Table 2.2
Flexible Design Components

Component	Summary
Multiple dialogues	Dialogue within duoethnography exists between inquirers, as well as between them and the texts they examine and create (e.g., photos). Dialogues create a multivoiced text and thus an increase in researcher imagination, perspective, and new meanings in relation to their co-construction of narratives of experience.
Insurrection of subjugated knowledge	In a recursive process, duoethnographers examine and attempt to surface deeply buried meanings as well as often subordinated counternarratives (narratives that run counter to those normalized within the dominant culture).
Critical collaboration	Working together, duoethnographers give each other feedback that seeks to decenter, deepen, and critique emergent meanings and understandings in quest of the researchers' transformation of those meanings.
Synergy of data collection and analysis	Data collection (story telling) and analysis (recursive critique and summary) are intertwined and mutually supportive. While the boundaries are blurred between them, they are each intended to deepen the other.
The signature of the reader	The signature of the reader refers to the meaning that the reader gives to the emergent text as she or he engages in it, thus changing its meaning in a transactional process.

in these textual transactions. Texts in duoethnographic studies are not limited to discussions between researchers. They also include more imaginative forms of data, such as cultural artifacts, poetry, photography, old report cards, yearbooks, film, and narrative representation or presentation of art. Researchers select their texts based on personal meaning and on the generativity of the dialogic potential of the text for their study. As they create a transaction between themselves and a variety of texts, they contribute their own interpretation—their own poetry—to a particular textual reading, creating a new text (Rosenblatt, 1994).

As they plan their studies, duoethnographers consider multiple collaborative ways to present and interrogate their narratives of experience. Although the specific stories they tell each other emerge in the research process, duoethnographers enter the process with a willingness to "mesh" their stories in ways that at times intersect, contradict, align with, clash with, and mutually inform one another. These stories serve multiple meanings, from exposing emic perspectives of structural narratives to transforming the mythologizing tendency of monologue to a hermeneutic stance.

Insurrection of Subjugated Knowledge

In terms of design, it can be helpful to view narratives as texts nested together dynamically. Duoethnographers select texts based on their capacity to reveal sociocultural and political contexts of experience—both within and between narratives. This process builds on Derrida's (1973) view of deconstruction as an interrogation of one's perception of the rhetorical framing or meaning of sociocultural texts. This interrogation promotes the "insurrection of subjugated knowledge" (Foucault, 1980, 2003). "Subjugated knowledge" refers to counternarratives, to local and subcultural ways of knowing and acting. This knowledge often reflects critical discourses (such as critical race theory, feminism, and queer theory) that express colonized, invalidated, or normalized unique cultural knowledge. Part of the research intent of duoethnography is the exploration and identification of counternarratives for their own sake: to reveal them to the researcher and reader. Their "insurrection" comes from their capacity to critique dominant discourses by juxtaposition. While the counternarratives arising in duoethnography may not decenter dominant discourses, they do offer counterpunctual narratives of experience to them. The research is designed to present disruptive texts.

Critical Collaboration

Duoethnographers have a responsibility not to let each other discuss and write about nonintersecting parallel story lines. While they might initially work in that way, the goal is to collaboratively create dialectics that promote complex thinking and deepen exploration of a problem of study—to build an imaginative capacity while working in the critical tension found in meaningful dialogue. To create this critical tension, duoethnographers listen actively to each other and to their data—specifically to the intersections and gaps within and between their narratives. This is an interrogation grounded in the uniqueness of the data, not in habitual and routine responses to them. For this engagement to occur, collaborators seek to provide each other with sufficient detail in their discussions. This generative detail is found in description, not explication; in questions, not definitive statements; and in story, not abstract understandings. It is also found in making connections between the study and authentic practice and real life, as researchers apply intersubjective knowledge from their research to their own narratives built around their research questions and problems of study. As mentioned earlier, Pinar's (1975) notion of currere provides a framework for grounding collaboration in the curriculum of everyday life. Currere positions the research process to promote researcher self-reflexivity and conscientization (Freire, 1970). Ideally, duoethnographers are willing to be questioned and challenged by their research partners.

Synergy of Data Collection and Analysis

In duoethnography, data collection (storytelling) and analysis (discussion) overlap in form and content. Stories are shared around specific themes that provide dialogic framing. This process is analytical, as duoethnographers listen to each other and make adaptations based on emerging questions and insights. Hackler's comments in his duoethnography with Breault and Bradley

(Breault et al., 2012) show this dynamic. Near the beginning of their paper, he makes this statement (in the third person) about himself as a boy growing up in Alberta:

> He knows that the only way to combat the boredom of a small prairie town is to escape into a world of creativity and invention, music and dance, truth or dare, hide and seek. Little does he realize that the path he has constructed before him will not run parallel to any he will cross in his life. (p. 120)

Here, Hackler is looking back from the present to the past, where he is looking forward to the present—a complex analytical act. When he writes, "Little does he realize...," he is folding an analytical reading of his narrative into its construction. Later, he discusses his narrative in relation to Breault's:

> I think what's amazing to me is every time I talk with you is that...our experiences growing up and I'm, you know, and I ended up being [*slight pause*] gay and you're not. [*Both laughing*]. And that's really great for me...because it brings another dimension of normality into my own identity. And I'm selfish to say that but it makes me feel that, "Yeah, you know, what if all those things [referring to experiences around elementary-age students] were also not necessarily because I was gay." 'Cause I didn't even know I was at that time. I just knew I liked fun things. I liked to laugh. I loved the youthfulness. (p. 120–121)

This text suggests that ways that Hackler had begun to see and label himself, possibly in relation to an internalization of societal norms, could be changed and he could restory his narrative. That insight—which becomes part of a new narrative of experience—involves both storytelling and analysis. He makes this restorying more explicit later in the paper: "I realized, for myself, for example, that the profession [of teaching] did not just find me, but rather that I found what I needed within the craft to evolve as a human being" (p. 132). This thought echoes themes that his co-researchers raise about the need for an authentic, lived, dynamic curriculum. Hackler's narrative shows how a teacher's (and now principal's) passion for an authentic curriculum is connected to a lived experience, his childhood back in Alberta.

Hackler then begins to deconstruct this theme of identity's contrasting with structural normativity by writing in the next line of his reflective essay:

Perhaps by squeezing my feet back into those once-worn shoes, I allowed myself another opportunity to question, through adult eyes, the norms and rituals of our classrooms and society that may or may not have had a positive or detrimental effect on who I am today. (p. 132)

In terms of design, it is important for the researcher to adopt a courageous and willing disposition to examine him- or herself and then fold understandings from that examination into the honest creation of new stories. To return to Hackler, his study encouraged him to change his practice:

As I sit here now, in a newly unwrapped "Principal's" chair for the first time in my life, I catch myself carefully constructing opportunities for my brand-new staff to explore, encouraging them to question and challenge the processes they have become a part of before, rather than after, they consider replicating such perspectives, behaviors, values, and paradigms in their daily professional and personal lives. (p. 133)

This passage also suggests the recursive nature of storytelling and analysis in a duoethnographic study. As the researchers engage in their dialogues and storytelling, the link between analysis and storytelling becomes deeper, more critical, more synergistic, and more revealing.

The Signature of the Reader

As a form of research grounded in lived experience, duoethnography seeks to have an impact on its readers. On one level, duoethnographies may be read as data. As duoethnographers research specific questions and problems, their investigations reveal findings, however uncertain and emergent, through a human and dialogic lens. Such findings are present, for example, in Krammer and Mangiardi's duoethnography on the hidden curriculum of writing they encountered as students growing up in Canada. The concept of the hidden curriculum involves the socialization process implicit in a more formal curriculum (e.g., that found in schools).

In the words of McLaren (1989, pp. 183–184, as cited in Krammer & Mangiardi, 2012, p. 42), the hidden curriculum "is part of the bureaucratic and managerial 'press' of the school—the combined forces by which students are induced to comply with the dominant ideologies and social practices related to authority, behavior and morality." Krammer and Mangiardi's study presents many "findings" about writing development—what was helpful and what was harmful to them as students. It can be argued that their powerful insights in their study would not have emerged in more conventional forms of research. Their study presents significant findings about writing, the hidden curriculum, and the lives of students that contribute greatly to this research field.

On another level, duoethnography creates the possibility of catalytic validity (Lather, 1986; Reason & Rowan, 1981), to readers to create meaning through a self-reflexive engagement in the reading. As Barone (1990) suggested, narratives enable readers to (con)spire or breathe (with) the text. In this sense, duoethnographers do not take their readers to be consumers of their findings. Rather, they consider their readers to make meaning in collaborative and dialogic ways. The benefit of duoethnography for readers is promoted by their co-collaboration in the research process. As participants, not consumers, they do not apply findings so much as engage the study as a catalyst for their own understanding and reflexivity. The dialogue involves the reader or hearer of a duoethnographic text as well as the researcher. As Roth (2005) stated, referring to Derrida (1985), "biography (ethnography)...comes to life only through the countersignature of the reader or hearer" (p. 11).

In terms of design, duoethnographers want to make their articulations and dialogues accessible to the reader. They want to invite the reader to be a part of the conversation. Instead of writing in an abstract, authoritative voice, duoethnographers usually write in the first person. A duoethnographer who writes in the third person, as in the example of Hackler above, makes his or her voice and identity present. To make their voices explicit, duoethnographers often write at least sections of their study in script format, alternating between speakers (Norris, 2008; Sawyer & Norris, 2009). They also describe experience using thick sensory detail. The goal is to present sufficient thick descriptive detail—showing, not telling—to allow the reader to engage imaginatively in and be altered by the conversations.

Conclusion

When duoethnographers consider design before beginning an inquiry, they think about how to ensure quality in their work. They consider ethics, issues of representation, issues of presentation (the use of multiple forms of literature and artifacts to build thick description), critical intersubjectivity, reflexivity and praxis, and the interconnection of their study with the real world (i.e., that it is not apolitical and value free). They think about how they are situated in relation to their topic and each other. They also think about the relationship between their topic and sociocultural and political contexts. Given their topic and their collaborative pairing, they consider dispositions—frames of mind—that will be helpful to collaborating in complex and demanding ways. They consider how they will illuminate each other's thinking; destabilize it; open it to multiple perspectives; break it free from clichés, biases, and narcissism; distinguish it from dominant narratives; and—together— restory and reconceptualize narrative views of experience. While design and methodology are intertwined, duoethnographers use their initial planning as a way to expand and benefit from the rhizomatic meshingness in working together.

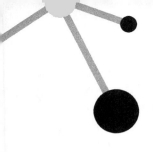

A LIVING METHOD

CHAPTER 2 USED the concept of the rhizome as a metaphor for duo-ethnographic research. This metaphor highlights the grounded, interconnected, crisscrossing, and symbiotic nature of the duoethnographic process. Elsewhere we have used jazz performance as a metaphor for duoethnography (Sawyer & Norris, 2009), foregrounding its emergent, dialogic, unpredictable, disciplined, and at times magical arrangements and motifs. In this chapter we discuss the emergent, creative, and dialogic methodology of duoethnography. We begin by presenting brief profiles of how two pairs and one trio of duoethnographers have constructed their studies, in order both to illustrate the process and highlight the diversity of approaches. After discussing some of the unique methodological characteristics of these three studies, we review central methodological considerations in duoethnography as practiced more than prescribed.

A Duoethnography on the Hidden Curriculum of School

In a recent duoethnography, Donna Krammer and Rosemarie Mangiardi (2012) investigated the hidden curriculum of their childhood public school experience in Canada. Given the increasingly politicized and regulatory nature of classroom curriculum, hidden curriculum is arguably one of the most important potential research topics right now, but one that is almost exclusively ignored. Krammer and Mangiardi state—quite provocatively—that they focused their research on "explor[ing] the content of this hidden curriculum, the way in which it has been structured within [our] own lives, and the influence that it has exerted on [our] concomitant decision to carry on with schooling at the doctoral level" (Personal correspondence to Sawyer and Norris, October 2010. Permission granted to cite). They explicitly examined the relationship between the internalization of the hidden curriculum as part of identity ("structured within [our] lives") and its impact as a socializing agent. As a learning case for methodology in duoethnography, their study highlights the uniqueness of the focus of inquiry, its frame for personal positionality (in relation to the hidden curriculum and its impact), and its potential space for self-awareness and reflexivity.

To explore this complex topic, they worked in a flexible, relational way. Because they lived in different parts of Canada, they communicated both in person and, more often, in writing as they constructed their study. They talked together, e-mailed each other, wrote, reflected, and talked again—circling around and digging into the details of their topic and how they had, and were, making meaning in relation to it. Dialoguing, they constructed a "rhythm of duoethnography—the back and forth, inside/outside, forward and backward movements that constitute duoethnographic exploration" (Krammer & Mangiardi, 2012, pp. 44–45). Furthermore, to recreate in writing the duoethnographic process they experienced and to bring the reader into the conversation as they present their study, they "deliberately disrupted the narrative at critical points in the telling to suggest the manner in which our conversation flowed from past to present, from narrative to analysis, and from self-reflection to commentary" (p. 45).

They selected currere as a conceptual framework for their study for its capacity to help researchers, curriculum theorists, and others promote a relational web to generate meaningful stories and personal engagement that leads to a reconceptualization of those stories: "As humans, we are both story-making beings and beings that are constituted through the act of story-telling—beings who in telling the tale, create and re-create the self, knowledge of the self, and knowledge of the world" (Krammer & Mangiardi, 2012, p. 43). Applying the concept of currere to duoethnography, they write:

> When researchers narrate lived experiences within the act of conversation, an *emergent* mode of communication, stories can develop organically, take unexpected turns, and assume intriguing contours in the back and forth movement between participants. In this way, conversation creates its own inimitable structure. By moving in, around, and beyond the topic at hand, it expands the possibility for understandings. Conversational form also embodies an epistemological truth: just as conversation is ongoing and open-ended, our understandings of past experiences will always be partial and open to revision. (p. 45)

This passage illustrates an intention to focus on a relational language-based approach "not so much to work with narrative, but to let narrative work with us" (Morris, 2002, as cited by Clandinin, 2011).

Working on their study over a number of months, Krammer and Mangiardi generated new insights and understandings within the tension of their flexible and open but intensely focused research approach. They describe this process with precision and clarity:

> In the conversational telling of our stories, we have had to reveal, reconstruct and reinterpret our memories of past events, examine the personal meanings arising from these interpretations and, then, allow those meanings to be transformed in the dialogical coupling and the juxtaposition of the telling. Throughout this process, unanticipated points of intersection—similar life experiences, or similar emotional responses under comparable circumstances –galvanized each teller to expand upon her story. As we unearthed long forgotten details, we burrowed deeper into these similarities until, eventually, we hit the bedrock of divergence. In this way we excavated each commonality to arrive at our inescapable human differences. Hence, the moments of connection and points of commonality ultimately drew our attention back to the critical points of contrast. On occasion we found ourselves manifestly in the here-and-now, discerning, and then analyzing, the language doing the telling—the words spoken, the gestures used, and the meanings implied. At other times,

we brought the past into the present, scrutinizing action and motives through the lenses of present-day perspectives in our efforts to explain pivotal events and understand their impact upon us. In this critically reflexive manner, we interrogated and reinterpreted our school stories. (Krammer & Mangiardi, 2012, p. 44)

In the phrase of Janet Miller (2011), they entered a "contingent community" for themselves, inviting their readers to join: "We hope that our readers enter into this conversation by revisiting their own school narratives, stories which, when juxtaposed with our own, may transform understandings and engender new insights" (Krammer & Mangiardi, 2012, p. 44).

A Duoethnography on Male Elementary Teachers

As mentioned in Chapter 2, Rick Breault, Raine Hackler, and Rebecca Bradley (2012) conducted a duoethnography on male teachers in public elementary schools. Their methodology is noteworthy for its emergent and responsive nature, its research questions focused both on identity and on methodology, and its creative use of online, electronic archiving to help promote trustworthiness and transparency in the study. They established their research questions pretty much at the beginning of their work together:

The guiding question for the duoethnography was, "What is the role of early male identity construction in negotiating one's place in a predominantly feminine work setting, such as the elementary school?" However, Rick was also curious as to the nature of the impact of the work environment on continuing construction or reconstruction of male teaching identity. Therefore, a second question emerged, "What is the nature of the influence of a predominantly female professional school culture on male identity construction?" (Breault et al., 2012, p. 118)

Breault and Hackler began working together as a pair in their study, but after a few meetings together they realized that as they collaborated, they were "heading toward a less than critical exploration" (Breault et al., 2012, p. 121), finding more commonalities than contrasts in their construction of their narratives, thus reinforcing rather than expanding their initial perspectives. As they put it:

> Reading the self-congratulatory tone of the early transcripts made us skeptical of our ability to honestly investigate our own development and ran the risk of what Norris and Sawyer have warned against—the possibility that the writers end up creating a "hero or victim saga." When this happens the storytellers tend to be, or at least appear to be, unlikely to undergo the transformation of perspective that underpins duoethnography. (p. 122)

To increase their study's critical focus, they added two research questions: "How can conversation partners avoid convergence—the mutual construction of a shared master narrative?" (p. 121) and "How do we more transparently represent the insights gained as a result of a complex interplay of dialogue, introspection, reflection and consideration of the research literature?" (p. 121). This simultaneous emphasis on the study of gender identity and on method made their use of reflection more explicit as a dedicated process in their work. The realization that they lacked a contrasting perspective led them to ask Rebecca Bradley to join their study. The study then went from being a duoethnography to a trioethnography.

With their researcher composition established based on its potential promotion of a critical intersubjective phenomenology (Schwandt, 2000), Breault, Hackler, and Bradley began working together in flexible ways, taping their conversations as they gathered in coffee shops, then transcribing these conversations and posting them onto their online site. From this site, visible to all, they reviewed all their conversations. As they tell it, "Embedded in th[eir] 'chat' [in coffeehouses] was a significant amount of ongoing reading, numerous e-mail questions and answers, personal introspection, conversations with family members and former students and post-conversation reflection" (Breault et al., 2012, p. 125). In their initial conversation they clarified the purpose of their project and of the duoethnographic process. They encountered an early dilemma between keeping their presented process transparent by including considerable unedited transcriptions of their conversations and keeping their study short, concise, and more thematically focused. They resolved this dilemma by placing the transcripts of all their conversations online: "To accomplish our goal of creating a naturalistic and transparent presentation we decided to use the hypertext capabilities of web-based publication." As part of this openness, they invited readers to discuss their conversations as part of their blog.

An additional step in their method was to organize their long conversations into more concise narratives. These narratives were followed by reflective summaries, in which they critically discussed key themes from their narratives. They stated:

> What started as an exploration of early influences on gender construction in male elementary teachers had moved toward the notion that there are experiences or early personality traits that might lead to becoming an effective elementary teacher. Had we stayed with our original conversation, Raine and Rick might have concluded that early, non-traditional gender conditions were largely responsible for their elementary teaching experiences. Rebecca's contribution challenged that conclusion. What seems more likely is that the personality traits and early life experiences of some individuals who choose elementary teaching contribute to becoming a certain kind of elementary teacher. (Breault et al., 2012, p. 131)

They situated themselves in these summaries very clearly in terms of personal accountability and self-reflexivity. Hackler, drawing from insights about personal agency and ways to (re)construct narrative views of life, considered how to work as a new principal with his teachers. To repeat a relevant quote from chapter 2 of this volume:

> As I sit here now, in a newly unwrapped "Principal's" chair for the first time in my life, I catch myself carefully constructing opportunities for my brand-new staff to explore, encouraging them to question and challenge the processes they have become a part of before, rather than after, they consider replicating such perspectives, behaviors, values, and paradigms in their daily professional and personal lives. (Breault et al., 2012, p. 133)

His is a complex thought that, as with Krammer and Mangiardi's study, echoes Pinar's (1975) notion of currere.

Bradley entered the study motivated to begin to understand conditions under which she questioned issues of elementary male teacher quality. Her own research question, then, acted as a counterpoint to that of Hackler and Breault. In possibly her first conversation with the two of them (found in her online transcript), she clarified her thoughts about her focus:

> But I feel that the majority of men that I have seen in teaching have been doing so in an effort to move towards administration. [*Raine nods in agreement.*] Regardless of what their background is or regardless of what it is they're teaching...I mean, that's sort of their goal and often times aren't strong in curriculum at all—which is a passion of mine...my opinion of other teachers is [often] based on that passion for curriculum and that drive toward the betterment of the curriculum and teaching practice. And I don't often see that. Unfortunately, [*continuing somewhat cautiously*] on multiple occasions I've seen men sitting back and reading the paper [*all laughing*] while their kids are doing work or they're out of the room and wanting me to watch their children while they go and run errands for administration. (Breault et al., 2012)

Implicit in this statement are a strong concern about male teachers in relation to the welfare of students and a possible recognition of certain gendered ways that male teachers interact with female teachers. Bradley's concern about the need for a dynamic and lived curriculum intersects with that of Breault and Hackler.

In relation to methodology, what appears to change for Bradley in the course of this study is not necessarily her views of teachers per se (the teachers she described are probably still reading those newspapers) but rather her self-understanding about this topic. In her reflective essay she writes: "These personal changes will invariably affect my professional practice, as I will view others, especially males, through a new and expanded lens. This growth opens doors to new learning experiences that would not have been possible in the past" (Breault et al., 2012, p. 133). She further states that this process

> challenged my somewhat negative perceptions of male elementary school teachers.... Our conversation...brought into relief that teachers are human beings, dynamic, complex and sometimes mysterious creatures. In society, we have the habit of categorizing or stereotyping individuals based on a few surface features and our past experience, which is often...limited. This practice unfortunately blinds us to

reality and possibilities. Had I encountered Rick or Raine in the classroom setting, prior to this research experience, I may have viewed them in a limited manner. Each small detail or action that confirmed my previous definition of a male elementary school teacher may have blinded me from seeing their pedagogical strengths and belief in a dynamic and lived curriculum. It may have constrained my desire to understand them, which in turn would have limited my opportunity and capacity towards learning and growth. (pp. 133–134)

With this thought she is describing a new vision of leadership—and of change—in her work with other teachers.

Breault in his reflective essay writes that this study has given him more questions than answers. Some of them intersect with those of Bradley and Hackler concerning teacher identity construction. He finds that their trioethnography, centered on identity, personal meaning, and authentic curriculum, provides a new and jarring lens with which to consider testing and accountability:

And now, as state and federal agencies continue to strip away teacher autonomy, standardize the process of learning and increasingly demand that success be defined in terms of measurable productivity, what will remain of the "feminine"? How will a feminized work culture respond to the imposition of such masculine, factory-like demands? Will the elementary school become a more comfortable place for career-driven men? (Breault et al., 2012, p. 135)

His concern, nested in the trio's nuanced narratives about hope and life, gives a new meaning to the abstract architecture of "measurable productivity" and "factory-like demands." As with Hackler and Bradley, these thoughts provide a new framework for engagement in the teaching profession, both for Breault personally and for the readers of the study. The group's study highlights a methodology for and of praxis.

A Duoethnography on the Null Curriculum of Sexuality

A third example that contributes to an understanding of methodology may be found in a duoethnography that we conducted on the hidden and null curriculum of sexual orientation (Norris &

Sawyer, 2004). This particular study stands out for the range of cultural artifacts that we examined both to revisit and to restory our perceptions of our lived histories. We started the study when we worked together in Washington State, with considerable initial discussion taking place on a three-hour drive through the rolling wheat fields of the Palouse in eastern Washington. We continued the study when we lived on opposite ends of the North American continent (Nova Scotia and Oregon) by juxtaposing face-to-face discussions at conferences (where we would also write together in conference and hotel rooms), formal presentations of the emerging texts at those conferences (thus creating a different form of dialogue, both with each other and with an interactive audience), and personal reflection and individual writing. Our discussions with numerous presentation audiences, who publicly critiqued our work, helped us to stay critical, as did our engagement with our arts-based cultural artifacts and our summary reflections in the text. With these different dialogues we attempted to destabilize and deepen our conceptions and reconceptions of our narratives.

For example, I, Rick, used the picture shown in figure 3.1 as a way to disrupt a narrative perception. A friend of mine took

Figure 3.1

that photo (with me present) about 20 years ago. In the study, I wrote:

> His photography presents disembodied images—with people often lost in a decontextualized or jarringly framed world. He gives a hand, a leg, or a foot a visual weight one would not normally assign to them and freezes them with all their blemishes. Commercial photography now has jarring edges and abrupt cuts, but 20 years ago, his style came from a vision, not a catalogue. His photos reveal animated surfaces, amorphous shadows, sharp edges, and radian light. (Norris & Sawyer, 2004, p. 144)

Over time, this photo has become increasingly abstract to me, with its images—disconnected legs, shadowy bodies, scraggly grass, amorphous background landmasses—now triggering a deeper meaning of the events of that day: A friend and I made a rare excursion to a beach. We took off our shirts and felt self-conscious. We saw beautiful people sharing a concrete, sensual moment. We felt disconnected, abstract, hopeful, ironic. When I see the photo now I love the sculptural quality of the bodies as well as the sense of disruption of a "normal" beach scene, as the jarring angles and sharp shadows frame passionate people. While the objects in my friend's pictures seem animated and infused with beauty, his shapes can be sinister and threatening, reflections of a world too confident to hide its biases.

Examining this photograph now (in conjunction with other artifacts included in the study, such as a participant journal and paintings), I see that we critiqued normal situations from the margins and broke their normative storylines. Examining a photograph out of its narrative context (its inclusion with other photos from the same photo session) can underscore the disruption of a storyline. Prikryl (2010) noted, "The way a photograph lops off a slice of reality, severing it from the narrative flow of time, is a seductive thing: it acts like a little hammer to the reflex in our brain that wants to tell stories" (p. 29). As my transaction with this picture (and the study) changes, I experience a new meaning: that many of my and my friends' youthful public acts of communication transcended our own situations to encompass a profound respect and hope for humanity. Our rebellion sought tolerance for all. As I write this sentence, my inquiry through this study continues.

As a result of Rick's use of photographs, I, Joe, photographed my former elementary and secondary schools as well as a local historical site. I wrote:

> The once fond memories were now etched with the scars of homophobia. I felt embarrassed taking the pictures as I was making explicit the null and hidden curriculums of my schooling. I wondered what the Sisters of Charity would say about my exposure. In choosing one photo, I decided to use the blurred image of my elementary/junior high school [left side of figure 3.2], rather that [sic] the clear one [right side of figure 3.2], to visually demonstrate the heteroglossia of the image. (Sawyer & Norris, 2010, p. 133)

In the same study, I introduced and analyzed prose I had written at an earlier date as a cultural artifact. I wrote:

> This following piece I wrote a number of years ago, shortly after it [a sex education meeting at a school] happened. My daughter was in grade eight and my wife and I attended the required invitational meeting to discuss the content of the "sex ed. program." I wrote it to think it through and to provide it as an example on how one can reflect on occurrences of daily life and relate them to educational issues for my preservice teachers. (Norris & Sawyer, 2004, pp. 151–152)

In this writing I explored the hostility that some parents expressed toward gay people. I then analyzed the meeting in terms of the null curriculum and my actions in it. As an artifact, the piece of writing I analyzed gave me the opportunity to revisit a product I made meaning with at an earlier point in my life: I reflected about

Figure 3.2

both the past and the present, in order to destabilize a theme with continuous resonance. In 2004 I wrote:

> But I found that I too practiced the null curriculum at home. I did not want to make a big deal over gay rights as I wished that my daughters would consider the lifestyle as a natural and legitimate lifestyle. I thought then that making it a big deal would reinforce the difference, possibly in a negative way. But my silence was problematic. (Norris & Sawyer, 2004, p. 148)

Although we conducted this study a number of years ago, how we have been situated in our study has remained uncertain and continued to change—as Joe's notion of "problematic silence" suggests. I, Rick, too have continued to revise my thinking about my narrative in relation to the study. When we conducted it, I thought that the photography and writing by my friends and me about gay rights them in some ways ironic and defensive; now I see them as embracive of and generous about the rights of all. In our study on the null curriculum of sexual orientation, we traced and redrew our emergent and changing meanings on this topic. By juxtaposing our stories, which expressed very different meanings while we lived in similar cultures at the same time, we looked back to look forward, were regressive to become progressive, and transformed our meanings related to this topic through our currere.

Methodological Considerations

While the three studies we just described are unique, they do share methodological themes. In each of these studies, the researchers generated their research focus as they built their relationship with each other. Their conceptual research tools, being relationally grounded, emerged from the nature of their work together. These tools included building and maintaining trust, constructing currere, constructing multiple meaningful dialogues, deconstructing experience, examining place-based pedagogy, and finding synergy in data collection (telling stories) and analysis.

Letting Narrative Inquiry (Re)author Our Lives

Through story duoethnographers enter their own narratives as critical sites of research and investigation. As they explore their narratives of experience, they examine not only patterns of cultural influences and inscriptions on their lives—ways of enculturation—but also their patterns of perception about such enculturation. This process reflects how narrative researchers "tend to begin with experience as lived and told in story" (Clandinin & Connelly, 2000, p. 128). As Clandinin and Connelly (2000) state, "[F]or narrative inquiry, it is more productive to begin with explorations of the phenomena of experience rather than in comparative analysis of various theoretical methodological frames" (p. 128). As we mentioned earlier, as duoethnographers engage in this process, they are not using narrative as a tool so much as letting narrative work with them. In this process, narrative itself becomes a frame for experiencing and interpreting one's own experience: It forms the ontology of narrative (Clandinin, 2011).

The ontology of narrative is unique to each pair of researchers. As duoethnographers work together, their methodology emerges. Common to all duoethnographers, however, is that their engagement in their inquiry is through their construction and sharing of stories to create experiential webs. Duoethnographers seek to interweave refracting narratives—interacting narratives that open new windows on experience. Paradoxically, duoethnographers seek both to enter these stories and to maintain a critical distance from them. As a metaphor, it is helpful to imagine these narrative webs as having temporal dimensions (past/present/future), situational/experiential dimensions, and conceptual dimensions. The researchers are simultaneously within and outside of them. In this storytelling process they seek to make sense of their stories in relation to meaningful patterns and themes. With an accompanying tension, duoethnographers construct, deconstruct, and reconceptualize their initially free-flowing story webs while maintaining an awareness of how they are situated to their narrative webs as they create them. It is as if they are watching this process in a mirror as they engage in it.

An example of two researchers taking a mirrored, reflective stance within narrative is found in Sonia Aujla-Bhullar and Kari Grain's (2012) duoethnography on mirror imaging of diversity experiences related to identity in cross-cultural initiatives. In their study, they placed the Canadian metanarrative of inclusion, prosperity, and democracy as a backdrop to their study of the intersections of their contrasting narratives and of the multiple codings and inscriptions of their identities. Aujla-Bhullar wrote:

> In the past year especially, working on research and looking at more complex ways of identifying myself, I've really come to be critical of how I define myself. The basis of how I would explain myself to someone else is, first and foremost, a woman of color, East-Indian, Indo-Canadian, second-generation immigrant born to parents who were born in India, in the northern region of Punjab. I'm also a teacher working with minority-identity students for the most part in a high-needs area of Calgary. And I am a graduate student studying forms of multiculturalism in Canada. The past few years, especially, I've really come to see my identity as being more complex than just referring to myself as an East-Indian, Canadian woman. (Aujla-Bhullar & Grain, 2012, pp. 200–201)

She continues:

> It feels like my identity is so embedded in every single thing that I do that there's no room to really step out of it and just…be me. But I think that's a part of being a minoritized person in Canada. We're very multicultural in the sense that it is almost synonymous with defining how we are Canadian, "proud" to be Canadian. We are all of these things that we are supposed to be proud of, but is it ever truly that? How does Canadian nationalism tie into the experiences of minority persons? (p. 202)

Grain then tells a contrasting story from her childhood:

> Mom's stories…always recalled narratives of being poor and not being able to speak English when she came to Canada…and how her parents built—just from working random construction jobs—how they built a whole life for their entire family. I guess in this way, my parents' stories have served as a sort of curriculum in my childhood and even now as an adult. The stories stay the same but they continue to reinvent themselves as I encounter new challenges and phases in my life. But I'll be clear on this: I wouldn't say that I've felt minoritized. I think there's a difference between feeling like a second-generation immigrant and feeling minoritized. I think that could be really different. (p. 204)

As may be seen in this short example, in their conversations they created *critical intersections* around differing experiences, leading to new stories and questions to pursue.

Another example of the decentering process of juxtaposed narratives is found in Darren Lund and Maryam Nabavi's (2008) duoethnography on social justice activism and issues of identity and racism in Canada. For Nabavi this cyclical process, much like that of Aujla-Bhullar and Grain, involved her exploration of her own narratives of power in relation to dominant discourses:

> I really think this was the most poignant experience in learning about how we can carelessly take advantage of our social locations and perpetuate power imbalances in our communities. I find that this understanding has created a tension that, at times, I do not know what to do with, from within my standing in the Eurocentric milieu of Canada. (Lund & Nabavi, 2008, p. 29)

She consciously promotes *self-reflexivity* by placing her personal narrative into the dominant narrative, using each to critique the other:

> I work to reclaim my multiple identities and become more aware of both the challenges and advantages that I carry. Finding a middle place to the extremes, as constructed by dominant rhetoric, is an ongoing process that I embrace through transformative approaches to social change and always being counter-hegemonic. (p. 27)

Both Aujla-Bhullar and Grain (2012) and Lund and Nabavi (2008) used duoethnography as a method to study the researchers—i.e., themselves—not the "other." In this process they not only undermine the Western notion of ethnography as a way to examine the other as subject or object (Latour, 1993) but also find a way to reconceptualize their own narratives. This process is rooted in each pair's trust of each other and in the uncertainty and unpredictability of telling stories (Western storytelling conventions notwithstanding). It echoes notions of the fluidity of identity and agency in cultural worlds (Holland, Lachicotte, Skinner, & Cain, 1998). In a duoethnography, researchers view their stories and histories (discourse and practice) not as indicators of essential themes but rather "as the media around which socially and historically positioned persons construct their subjectivities in practice" (Holland et al., 1998, p. 32). Through dialogue, they construct their emergent subjectivity in practice in relation to the narratives they construct. Duoethnography views the self as embedded in (social) practice and as itself a kind of practice.

Constructing Heteroglossia

As duoethnographers use language together, they promote a process of heteroglossia (Bakhtin, 1981)—of meaning generation found within dynamic texts. As Holquist (1981) explained,

> Heteroglossia is... that which insures the primacy of context over text. At any given time, in any given place, there will be a set of conditions—social, historical, meteorological, physiological—that will insure that a word uttered in that place and at that time will have a meaning different than it would have under any other conditions; all utterances are heteroglot in that they are functions of a matrix of forces practically impossible to recoup, and therefore impossible to resolve. Heteroglossia is as close a conceptualization as is possible of that locus where centripetal and centrifugal forces collide. (p. 428)

Through language and their use of diverse cultural artifacts (e.g., soap, popular music, lesson plans, newspaper clippings, math problems) duoethnographers create heteroglossia as they juxtapose and discuss pertinent events, time periods, understandings, questions, and cultural artifacts. Cultural artifacts have notably played a rich and varied role in promoting dialectical tension in duoethnographies. Artifacts have served many purposes. They tell us about our times, experiences, and history—now and in the past. They create dialogues between us and them through close and critical readings. They allow us to see our perceptions of an event and thus to reconstruct that perception. And they provide rich description for a duoethnography's researchers and readers.

A glance at a few of the artifacts used in duoethnographic studies suggests their usefulness. Tonda Liggett and I, Rick (Sawyer & Liggett, 2012), did a study of colonialism in our teaching. To examine our currere, we pulled out old copies of our personal high school yearbooks and explored forgotten and new meanings of, among other things, the sexism inherent in their photos and in their captions of photos of student clubs. Exploring current forms of privilege and entitlement in the southern United States, Francyne Huckaby and Molly Weinburgh (2012) examined the old Confederate national anthem, "Dixie." As a song they had known all their lives, their chosen artifact was deeply embodied. And in their exploration of their curriculum of beauty, Morna McDermott McNulty and Nancy Rankie Shelton (Shelton & McDermott, 2012) examined their high school diaries and photographs from various times in their lives. They explored how class—as signified by a bar of soap and a shampoo bottle—influenced their construction of beauty. While the soap in their study is an imagined artifact, its sensory evocation is immediately present. These duoethnographers selected and included artifacts in their studies based on their evocative power. They moved their artifacts through time and interrogated them from the vantage point of the past, the present, and the transaction between the two. With imagination, these researchers conceptualized a new narrative from the promise and resistance of their artifacts.

Creating meaning and transacting with an artifact (Rosenblatt, 1994), duoethnographers engage in an arts-based phenomenology (Noe, 2000). As Noe (2000) stated: "The work of some artists can teach us about perceptual consciousness by furnishing us with the opportunity to have a special kind of reflective experience. In this way, art can be a tool for phenomenological investigation" (Noe, 2000, p. 123). Maxine Greene (2001) notes that art is "left contingent, partial, wonderfully open and incomplete for the reader to develop and to realize, to be compelled to read what they have never done so . . ., to look at pictures that they have never seen, to attend to music they have heard only in the distance if at all" (p. viii).

It is within this "contingent, partial, wonderfully open" space that visual phenomenology reframes our awareness of how we interact with visual experience. Building on the notion of meaning creation as an act of intersubjective interpretation (Schwandt, 2000), visual phenomenology serves as an interactive tool for the investigation of visual and environmental spaces by emphasizing the unstable transaction (Eisner, 1991) that takes place between a viewer and art. This transaction promotes an intertextual construction of meaning, because the meaning of the art is not contained within the form itself (Sullivan, 2005) but rather emerges through a transaction between the viewer and the artwork (Eisner, 1991). Sullivan (2005) describes this process:

> Meaning is not within a form itself, say a person, painting, or a poem, but exists within a network of social relations and discourse. This interpretive landscape of "intertextuality" serves as the means by which meanings become distributed and debated. (p. 43)

This transaction contributes to the methodology of duoethnography in multiple ways. First, the meaning of the artifact in a study is never fixed but rather is always in flux and coming into meaning. Second, the meanings themselves, then, are polysemous—marked by a multiplicity of meaning. This tension between the artwork and the viewer is a way that art promotes the "subversive imagination" (Becker, 1994) of the viewer, creating a context for reflexivity and praxis. Third, "visual art . . . can . . . effectively be used to challenge, dislodge, and transform outdated beliefs and stereotypes. . . . The arts always retain oppositional capabilities" (Leavy, 2009, p. 216). As Leavy (2009) stated:

> The kind of dialogue promoted by arts-based practices is predicated on *evoking meanings*, not denoting them. In other words, although qualitative research typically claims to be inductive by design, it often falls short with preconceived language, code categories, and guiding assumptions creeping into the process, often more than we may realize. Arts-based practices lend themselves to *inductive* research designs. (p. 14)

Returning to the example of the high school yearbook photos Tonda and I discussed and examined, we attempted to use art as a means of "decolonizing the imagination" (Becker, 1994, p. 111). We found that given photography's frequent lack of internal references, photos disrupted the narrative within which they were taken and which they were intended to support. Taken out of context, they forced us to examine how we made meaning at the time they were taken. We then, to refer again to Leavy's (2009) discussion of arts-based research, examined the "institutional context with various restraints, norms, pressures, and so forth influencing [their] production and circulation, as well as the value system within which [they were] judged" (p. 216). We used our discussion of those factors as a lens through which to view our current narratives of colonialism inherent in our teaching. Through our dialogue, we attempted to reunify the photos in new narratives, as we sought new narrative unities. However, because we witnessed the contingent nature of our social construction of meaning around those images (Miller, 2011), this unity remains uncertain.

The use of artifacts in duoethnography draws from Clandinin and Connelly's (2000) work with field texts in narrative. Field texts provide inquirers with a means to "move back and forth between full involvement with participants and distance from them" (Clandinin & Connelly, 2000, p. 80). As they noted, "Field notes, photographs, students' written work, teachers' planning notes are all field texts that help us step out into cool observation of events remembered within a loving glow" (p. 83). They "assist memory to fill in the richness, nuance, and intricacy of the lived stories and the landscape" (p. 80). Thus inquirers' use of field notes is rooted in their need to begin to examine and make sense of how they are situated in relation to the nature of the inquiry, one in which they themselves are relationally embedded. Field notes are a means to help inquirers begin to explore in writing and through other media the relational and spatial tension between themselves and their inquiry setting, participants, or topic. Duoethnographers attempt to access and promote this relational and spatial tension as a dialectical frame in their studies. In duoethnography, this tension is grounded in inquirers' stories about their own lives.

An Analytical Lens: Currere and Place-Conscious Education

As we have illustrated, data generation (storytelling) and analysis occur simultaneously in duoethnography. The data analysis process unfolds from researchers' patterns of meaning making within their (re)creation of experience in their stories. Kim (2006) describes this process in relation to narrative analysis as a configuring of stories. In this process, "the researcher extracts an emerging theme from the fullness of lived experiences presented in the data themselves and configures stories making a range of disconnected research elements coherent so that the story can appeal to the reader's understanding and imagination" (Kim, 2006, p. 5, as cited in Leavy, 2009, p. 28). Good analysis involves dialogue with another and extracting the story from the described experience (Ellis, 2004). In duoethnography, the extraction of theme is framed by currere and place-conscious pedagogy. Place-conscious education or pedagogy provides duoethnography with a defined focus on different dimensions of place. It highlights the communicative aspects of place, thus giving the concept of place an active role in the research process.

Place-conscious theory presents duoethnographers with a context for telling and analyzing stories around the articulation of placed-based meaning. Greenwood (2009) characterizes place-conscious education as follows:

> Place-conscious education...can potentially challenge learners to consider where they are, how they got there, and to examine the tensions between different cultural groups' inhabitation across time. In every case, in every place, this would mean listening for the voice of Native survivance, with an ear for learning from the relationship between Indigenous ways of knowing and local and global narratives of colonization and contestation. In the context of this remembering, place-consciousness also suggests a reassessment of all current inhabitants' relationships with land and people, near and far, now and in the future. (pp. 3–4)

As can be seen from this quotation, place-conscious theory contains multiple dimensions that apply to duoethnography. Duoethnographers work in multiple contingent intersections of identity, culture, history, narrative, politics, and reflexivity. On a conceptual level, these contexts shift according to their location and researchers' relative positionality.

How place reflects culture and self-perception is illustrated by a story Tonda Liggett told about how the pedagogy of place led her to a personal awakening:

> I spent the semester abroad living and traveling throughout Kenya, walking among the Maasai tribe on the edge of the Rift Valley as one of a small group of college students venturing out to learn about orienteering, biology, history, and the culture of this new environment. No longer could I rely on my background information to give me an advantage

in this learning situation. In small villages, people came up to me and asked questions about my life, where I was from, what the United States was like. In answering, I realized I was pressed to respond for all whites—a representative of my race.... In speaking for my country I was speaking as racial minorities are often called upon to do in classrooms across the nation.... This semester abroad, I discovered, marked the beginning of an awareness about colonization, its legacy, and the ways that post/decolonization was apparent not only in my own personal identity, but also in my teaching and in language arts curriculum. (Sawyer & Liggett, 2012, pp. 74–75)

Tonda did not assume a tourist mentality and use her memory of her visit to Kenya as an opportunity to reminisce and reinscribe her home country values. Instead, considering how that trip "marked the beginning of an awareness about colonialism," (Sawyer & Liggett, 2012, p. 75) she began a new critical journey at home, where she began to deconstruct her own stance toward culture, power, and personal identity.

Complementing place-conscious education, the notion of currere (Pinar, 1975; Pinar, Reynolds, Slattery, & Taubman, 1995) contributes a conceptual framework in duoethnography for promoting researchers' multidimensional narrative engagement, reconceptualization, and self-reflexivity. Currere presents a pedagogical lens for researchers (and nonresearchers) to explore and experience dialectical dimensions of one's lived curriculum. As a research frame, currere allows researchers to ground the details and events of their lives within larger narrative meanings. It is currere's emphasis on narrative (examining life holistically over time from multiple perspectives) that engages the dialogic imagination. Tracking how one comes to believe something ideally leads to a change in the corresponding beliefs, a reconceptualization of self and world. Krammer and Mangiardi (2012) stress that "currere...is not an exercise in solipsism; it must communicate 'the individual's lived experience as it is socially located, politically positioned, and discursively formed' [Pinar, 1995, p. 416]" (p. 43). The emphasis is on self-reconceptualization, not mere recollection (Krammer & Mangiardi, 2012). Duoethnographers thus use currere as a tool to engage and examine the ever changing process of giving meaning to phenomena in their lives.

An example of how place-conscious education and currere can frame storytelling and data analysis is found in the previously referenced study by Aujla-Bhullar and Grain (2012) on mirror imaging of diversity experiences related to identity in cross-cultural initiatives. In this excerpt, Aujla-Bhullar writes about playing soccer as a child with European Canadians on a field in Alberta:

> One of the most poignant memories for me was when I was...playing soccer or some sort of a game. And there was this East Indian woman that walked by. She looked like my grandma, you know, with the outfit, the pyjami or salwar suit and with a covered head and everything. And she walked past our game and...my friends started talking among themselves and saying, "I'm not a racist or anything but there's getting to be way too many of them here." And I remember someone else saying, "Yeah, that's what my mom says too. We don't have anything against them, but there are too many here...."
>
> And I think that really...it hurt me, but at the same time I was scared because I didn't want them to see me like that. I was embarrassed because that was me and I didn't know what to say. I remember going home and just wondering, "Are there too many of us? Am I one of the 'us'?" They didn't look at me like that, but why were they playing with me? I started thinking of that and I still remember that day very clearly. And I'm still feeling that sick feeling to my stomach that I didn't say anything. I should've stuck up for myself and...and my culture. I'm so embarrassed about that still. I reflect on it now and see how it affected the construction of my identity; how others perceived me and how I perceived myself. (pp. 204–205)

Aujla-Bhullar's comment that "I was embarrassed because that was me" illustrates the complexity of interconnected temporal, personal, and cultural dynamics of currere. Grain adds a dialogic dimension to this emerging currere when she follows with her own childhood story about a good friend she invited to her house who then unexpectedly proceeded to make racist comments:

> I was sick to my stomach and I got up and walked inside. And I felt paralyzed and I didn't know to tell him to "get the

f— out of my house." I was so angry and upset especially since it was in…my territory. It was in my family's space. And, this is a spiritual space in a sense; it's a space where you're secure and where you practice your own beliefs and live out your own values. And everything he had just said and done went against that value system. I felt so violated and I didn't know how to ask him to leave because I was scared. I was scared he would yell at me or embarrass me in front of all of our friends. That incident has haunted me because I was a coward. I often think, "Man, I wish he would've pulled that now." (pp. 205–206)

These stories invite individual cultural meanings and values to clash with larger cultural meanings and values. They are framed by place and engaged through currere.

Another, very different perception of place is found in the duoethnography that Deborah Ceglowski (in press) conducted with children living in poverty in the United States. She notes in her study that she encountered challenges using duoethnography as a method with seventh-grade children, but she also mentions that she turned to it out of frustration in attempting to access children's lived worlds with more conventional forms of qualitative research. A few quotes from her study suggest that her young co-researchers had a uniquely located sense of place. Ceglowski writes:

Other children joined in with stories of things they would like but their families could not afford. These included: a house, new clothes, healthy food, and involvement in more school activities[,] including money for book fairs, participating in football, basketball, volleyball (and all other sports teams) and attending monthly school sponsored skating ($12/month plus money for snacks) and bowling. With the exception of two children, none of the children had access to a computer at home and this made it difficult for them to complete certain homework assignments.

Malachi stated, "We don't buy like meat or chicken." The other children added food items they would like to have but their families could not afford: tomatoes, celery, carrots, kiwi, apples, biscuits, and fish. The children also stated that they would like to eat in restaurants but rarely had. Jordan, aged 11, who wanted to work at Olive Garden when he grew up[,] told us that he had "only ate out in my whole life three times."

Mindy, a nine-year-old girl sitting with Dominique, said that her family went to Goodwill and that her 12-year-old sister wanted "to buy a lot of things but mom says that she can't." Mindy also said that she and her sisters received hand-me-down clothing from her cousins but that "I can't fit them."

While Ceglowski did not focus on place as an explicit analytical frame in her study, these children's sense of place—boundaries delineated by immediate circumstances and television fantasies, families offering love but suffering stress—needs to be understood as they lived it.

Ceglowski then interweaves her own childhood memories into her co-researchers' stories, finding both similarities and differences. In terms of currere, she moves her sense of the past into her future experience when she states:

The children and I knew what it meant to go grocery shopping and not buy all the foods we wanted to get; my children, like those I spoke with, had not received the birthday and Christmas gifts they wished for and often could not participate in school activities because there was not enough money to finance them. We had shopped at second-hand stores and knew that often there was not enough money to buy all the things that we would like to have.

She adds:

Though my income dramatically changed when I became a single parent, I was working at a University and earning a much higher salary than the children's mothers. My position included health care coverage; an area that the children did not discuss but undoubtedly was missing in several of the families.

Her recognition of this key difference—that she was receiving something that her co-researchers did not even discuss—is a highly compelling self-realization about self, self-privilege, and positionality. In place-conscious education self-realization is a form of—to borrow Greenwood's (2009) word—reassessment.

And in both currere and duoethnography, it is a context for reconceptualization and self-reflexivity.

Ceglowski's study illustrates how method in duoethnography both is about experience and is itself experience. This experiential engagement helps researchers delineate their central epistemological assumptions in relation to local meanings as well as larger global discourses such as colonialism. Greenwood (2009) suggests that grand narratives are our "[c]ultural assumptions or 'root metaphors' like individualism, anthropocentrism, and faith in progress...common to the dominant culture (Bowers, 1997)...[and] now common to a commodified American landscape (Kunstler, 1993)" (p. 3). Using place-conscious education ideally promotes a sense of consciousness raising (Freire, 1970), thus providing a context for researchers to (re)inscribe their "root metaphors" within surrounding dominant discourses.

Conclusion

Creating rhizomatic progressions, duoethnographers examine lived experience from multiple dialogic perspectives. They generate experiential details that animate the place-situated nature of their stories and how they are positioned in relation to sociopolitical and cultural contexts on local and more global levels. The goal of duoethnography is not just to promote researcher self-knowledge. It is also to promote researcher exposure or reformation of the dynamic between personal meanings and their sociocultural inscriptions. Duoethnographic methodology contrasts with that of qualitative research that intends for researchers to "uncover" conclusions and discover "findings." Although such findings exist in a relationship between the data and the researcher, they usually seek to articulate the experience of participants (those who are studied) but not explicitly the experience of the researcher. The participants in duoethnography, however, are the researchers, and the intent of its methodology is to create an ontological paradigm for researchers to understand, engage in, and transform perceptions of experience. As noted earlier, researchers in duoethnography do not work with narrative, it works with them. Duoethnographers do not so much "find" experience, as story and restory it—around the human voice—within cultural contexts.

WRITING DUOETHNOGRAPHIES

WHAT DISTINGUISHES DUOETHNOGRAPHY WRITING from most other collaborative forms of research writing is that the voice of each researcher (and researched) is made explicit. Rather than combining their ideas into the unified voice of a single narrator, duoethnographers create polyocular (Maruyama, 2004) or polyvocal texts presenting multiple perspectives on a phenomenon, avoiding the metanarrative of a singular point of view. Most often written in a theatrical script format, duoethnographies permit readers to witness two or more people both in conversation and thinking about their conversation. With multiple voices fracturing the notion of a solitary whole, readers are less likely to align with a single narrator. This structure frees readers to pick and choose aspects from each duoethnographer that they deem relevant and to use their choices to reinform their own beliefs and behaviors. Not only do readers witness dialogic conversations, they are invited into them.

Norris (1989) found that this was the case when audiences viewed performances of a play that consisted of a series of vignettes on the topic of growing up. Audience members recollected their own lives during the intermission and after the performance. The performed stories evoked their own dormant stories, and they shared these with one another as they juxtaposed their experiences with what was performed. The format created a "third space" (Bhabha, 1994) in which different perspectives could converse and reinform one another. Duoethnographic studies are written in a similar manner. The aim is to create a space in which readers can reflect on their own lives as they witness similarities to and differences from the lives of others (Lévinas, 1984). The research process is dialogic, and duoethnographies are written in a format that continues the dialogic conversation, albeit with future readers.

A duoethnography is written as a conversation between two or more people, and being reported in a linear fashion gives it the appearance of occurring over a short time. This, however, is not the case. Much background work takes place in generating the data for the text. Notes or transcripts from personal conversations, internet correspondence, personal writings, and collaborative writing sessions all contribute to the final piece, and more data emerge from the writing process itself. Discussions of theoretical literature often are found within the conversations and become a natural part of the duoethnography.

Writing duoethnographies, then, involves much editing and resequencing to obtain a natural flow that will be both coherent and interesting for readers. The process is not dissimilar to the scripting of the film *My Dinner with Andre* (Malle, 1981). A review by Ebert (1999) reported:

> The conversation that flows so spontaneously between Andre Gregory and Wallace Shawn was carefully scripted. "They taped their conversations two or three times a week for three months," Pauline Kael writes, "and then Shawn worked for a year shaping the material into a script, in which they play comic distillations of aspects of themselves."

In duoethnography, then, writing is simultaneously a form of data generation, data interpretation, and data dissemination, in not so nearly as linear a fashion as the finished text suggests.

Admittedly, duoethnographies are constructed texts, but so are all forms of research dissemination. Statistical data are organized into tables, interviews are reduced to excerpts, and observations are reorganized into themes. As Richardson (1990) claimed,

> Even the shape of the conventional research report reveals a narratively driven subtext: theory (literature review) is the

past or the (researcher's) cause for the present study (the hypothesis being tested), which will lead to the future—findings and implications (for the researcher, the researched, and science). Narrative structures, therefore, are preoperative regardless of whether one is writing primarily in the narrative or logico-scientific mode. (p. 13)

The narrative structure of a duoethnography is a scripted conversation with theory blended into the thick description of how two or more individuals experienced the phenomenon they are investigating. Such a conversational style can make duoethnographies more accessible to people who are not conversant with traditional academic material, addressing Leavy's (2009) call for research that reaches a wider audience. An unsolicited comment from Ryan Harris, an editorial assistant who worked on an earlier duoethnography manuscript (Norris, Sawyer, & Lund, 2012), supports this claim:

First though, I want to say that I found this book really engaging! There are some fascinating, deep conversations here that really took me by surprise. The little moments scattered throughout where one author would come across a new insight or understanding based on a previous point from his or her coauthor were very natural—and the organic nature of these conversations was really, really nice to see. What I'm trying to say is, as a layman I really enjoyed what I read! (personal communication, August 28, 2012, permission granted to cite)

Narrative forms such as duoethnography, then, have the potential of being accessible to a wider population than does conventional scholarship. As Haven (2007) claimed, "people are eager for stories. Not dissertations. Not lectures. Not informative essays. For stories" (p. 8). Much research is steeped in codes unique to its genre, excluding people who are unfamiliar with the particular stylistic traditions. Stories and scripts can be written in a manner that elucidates insights on the phenomenon under investigation that have value to both academic and nonacademic populations. Such a style does not make them less scholarly, merely more accessible. Ideally duoethnographers delicately balance their texts so that multiple populations can comprehend them.

Writing duoethnographies is both a research process (form of data generation) and a research product (dissemination).

Duoethnographers employ writing as one way of simultaneously articulating and evoking memories and analyzing them as they coemerge both individually and collaboratively. The writing, consequently, is a research artifact, a research process, and a research product, as much of what is written will find its way into the final paper.

The process is usually a combination of solo and collaborative acts. One duoethnographer can take the lead by starting the written conversation and then physically or electronically passing it on to her or his fellow duoethnographer. The partner responds, providing insights and reporting stories evoked by the first piece. However, the linearity witnessed in the final product may not be the same as the structure(s) produced as the writing emerges. Each response may be given at the end of the other duoethnographer's draft or interjected into the existing pieces as further insights or connections are made. Norris and Cope Watson (2011) used text in different colors to indicate their updates as they e-mailed their responses back and forth. Once a draft was read, the reader changed the font to black and inserted responses in a different color. Each draft was numbered, providing a paper trail for each part of the written conversation.

While each duoethnographer's voice is made distinct in a duoethnography, editing by either party can and does take place. Minor phrasing and vocabulary suggestions are made and often accepted. These take place behind the scenes and are not made transparent. The designated voice of one duoethnographer can contain editorial suggestions and even content from the research partner.

While small accepted changes are not explicitly reported, disagreements and points of contention are. There is no attempt to resolve conflict and achieve a synthesis. In duoethnographies, when different points of view are expressed, both are written. The rationale is that by providing multiple perspectives to readers, the researchers enable readers to form their own, unique syntheses. In so doing, writers of duoethnographies invite readers to form their own conclusions. The writers may reflect on their own meanings, but they do not attempt to impose them on the general population. In Norris and Cope Watson (2011), Cope Watson challenges Norris on a comment that he made:

Georgann: I wonder if you asked your siblings if they would have the same memories about this. I wonder about the hidden or null curriculum of gender and education. Especially since it was the 60s and 70s. The circulating discourse was that women went to university to get their MRS [i.e., to become a "Mrs."] or to become more interesting partners for their future husbands. Sorry Joe, just pushing a bit. I always wonder how men can speak for women, when their experiences are inherently different.

Norris accepts the challenge and responds:

Joe: Push! I know what my parents expected and that is what I reported above. Read closely. I never spoke for any of my siblings, male or female. I related what my parents communicated to all their children. Ideology and generalized beliefs can also "frame" the way we perceive things. All [my] sisters work and from what they have told me over the years...this takes up a large part of who they are.

Rather than resolving an issue for readers, duoethnographers include this type of interaction, demonstrating the complexity of telling one's stories and implicitly inviting readers to draw their own conclusions. This type of exchange is built on reader response theory (Rosenblatt, 1978), which emphasizes understanding the transaction between the reader and the text. Rather than explaining the meanings individuals give to life experiences, in duoethnographies researchers portray two or more people in search of meaning, thereby implicitly inviting the readers to join the quest.

Since the first duoethnography (Norris & Sawyer, 2004), a number of other teams have conducted duoethnographies, which follow the basic tenets while varying in writing style and approach. Cunningham (1988), in his discussion of dialogic research, claimed, "At its simplest, there is no group process to attend to, only the interpersonal relationship of two people" (p. 164). In duoethnographies the data and writing coemerge from the conversation of two or more people; there can be no a priori prescription of text. However, there can be a range of possibilities. Building on the duoethnographic research thus far, we describe below an array of styles to provide those planning to write duoethnographies with choices about how to structure their writing—something that was not available to the pioneers of this genre.

Degree of Conversation or Tandem Writing

The first duoethnography (Norris & Sawyer, 2004) could be considered to use more of a tandem style of writing. After a prologue introducing the study, Norris tells part of his life story, followed by Sawyer telling part of his. This continues for a sequence of ten "scenes." Although Sawyer and Norris build in segues and refer to each other's stories, the chapter is more of a chronological collage than a conversation. Each piece remains separate, with little interaction between the duoethnographers. Sawyer and Liggett (2012) follow a similar structure but the authors each comment more on how the other influenced their thinking. Sawyer states, "Reading Tonda's section, I began to wonder how my own cultural indoctrination played out in a deeper, more layered and contradictory way" (Sawyer & Liggett, 2012, p. 181). Krammer and Mangiardi (2012) use a tandem structure similar to that of the first duoethnography, but rather that labeling their sections scenes, they call them "lessons." They also refer to each other's stories, making certain aspects explicit for themselves and their readers. For example, Kammer states, "You remind us, Rosie, that ideological critique requires evenhandedness" (p. 56).

The second duoethnography (Norris & Greenlaw, 2005, later published as Norris & Greenlaw, 2012) opens with a question posed by Greenlaw which Norris answers and then a questions posed by Norris, which Greenlaw answers. The writing style is that of a conversation between two people. As Greenlaw and Norris tell their stories, they discuss them with each other. The fourth wall (Brecht, 1957) is removed and the reader witnesses a discussion. There is continual interplay between the duoethnographers. Greenlaw asks, "So tell me, Joe, when did you discover the mysterious pleasure of being a writer?" (p. 92) and Norris responds. As the conversation unfolds, they explicitly juxtapose their differences. Norris states, " Unlike you, Jim, I had no early exemplars" (p. 95). There are no headings in this duoethnography. The conversation flows like a theatrical script. Even the tables that juxtapose the researchers' positions are introduced in a conversational format.

Joe: So, for a seventh inning stretch, let's recap our insights thus far, itemizing our writing similarities and differences through the following.

Nabavi and Lund's (2012) duoethnography parallels Norris and Greenlaw's structure. There is no division into sections, and as

Joe	Jim
Wrote at the request of others Extrinsic motivation	Wrote because the stories of others beckoned him Intrinsic motivation
Found writing painful	Found a pleasure in writing
Evoked by dreams	Evoked by dreams
Wrote for an audience Wrote for self Wrote for needs of others	Wrote for an audience Wrote for self From exemplars
Articulation of ideas, not polish	Enjoyed the craft of writing
Inspired by television	Inspired by books

(p. 103)

the stories are reported, each duoethnographer makes comments in a manner that could occur in a natural conversation. Nabavi comments to Lund, "While your story makes me cringe with the feelings of oppression your classmate must have felt..." (p. 185). Duoethnographies by Aujla-Bhullar and Grain (2012), LeFevre and Sawyer (2012), and Sitter and Hall (2012) follow a similar structure.

A third major structure could be considered of hybrid of the previous two. Huckaby and Weinburgh (2012) wrote their duoethnography in a fashion that has a conversational style but is framed with discrete segments. Prior to starting their scripted conversation, they begin with an abstract, and they include section titles that bring into focus the points they wish to make. Titles such as "Our Avenue of Transformation" (p. 159) and "Restorative Steps" (p. 175) shift the flow of the conversation, and they conclude each section with a voice of unison that they call "Sojourners." McLellan and Sader (2012) write in a similar fashion, as do Shelton and McDermott (2012).

Regardless of the structure, duoethnographers always make explicit who is saying what. If a tandem, heading approach is used, the writer/speaker's name is included in the heading. The degree of conversation is variable, but the articulation of whose voice is whose is not. However, this and other structural components of a duoethnography need not be predetermined; being aware of a few possibilities can enable writers to use some basic structures and adapt their content to the form they deem most appropriate.

As studies that employ duoethnography multiply, we anticipate that other variations in writing style will be created.

Headings

As previously indicated, the use of headings is one way of organizing the data of a duoethnography. This writing structure can bring into focus salient points, which can be themes, chronological progressions, or both.

Headings can be employed in a variety of ways to demarcate shifts of focus in the data and the writing. In Norris and Sawyer (2004), Norris organized his section chronologically, often using the word "becoming" in the section titles. This showed the progression of his informal curriculum/currere. Sawyer, however, gave his vignettes titles like "*In Your Pocket*," referring to the title of a book written by a friend (Fisher, 1996). Duoethnographic partners need not follow the same style. Difference is heralded in this methodology, and that goes beyond content and perspective to include modes of presentation. Krammer and Mangiardi (2012) start each section with the term "Lesson," and as each section advances chronologically, a theme embedded in the current story is extracted and made part of the title. For example, the section entitled "Lesson 8: No Autonomy Allowed" (p. 60) discusses the hidden curriculum of voice, or lack thereof, in the classroom. Breault, Hackler, and Bradley (2012) also use chronological titles; however, their titles mark more the chronological shifts in their study than they do each researcher's currere regarding their beliefs about male elementary teachers. Shelton and McDermott (2012) take a more thematic approach to their headings, providing titles like "Beauty and the Generation Gap" (p. 234) and "Beauty as Power" (p. 239).

Headings, however, like all writing choices, are problematic. In duoethnographies, choices regarding headings have epistemological, ontological, and axiological implications, because they frame the manner of knowing, the experience of reading, and the degree of authority of the authors. While headings can make certain constructs explicit, they have a tendency of "telling," not "showing." Reason and Hawkins (1988) claim that there are two basic forms of presenting data, "explanation" and "expression." The former tells, and the latter shows. Providing abstract headings can disrupt the flow and predispose the reader to reading through the researchers' frames. Omitting such frames permits more openness to interpretation, enabling readers to determine how they wish to "conspire" (Barone, 1990) with the text.

Introductions

Duoethnographies can but need not begin with an explicit introduction or some form of abstract. While such introductions can and often do serve a useful purpose, they, like headings, frame the writing that will follow. They bring a focus to the topic under investigation by discussing background literature, giving a summary of the research process, or both. Most often, these types of introductions are set apart stylistically from the scripted conversation and written in traditional prose with an implied composite narrator. Breault, Hackler, and Bradley (2012) begin with a summary followed by a short literature review before bringing in specific voices. The introduction tends to be less personal than the remaining sections. Shelton and McDermott (2012) make their introduction more personal through the use of the pronoun "we." In so doing, they indicate their personal involvement with the phenomenon while they discuss it.

The opening of Norris and Greenlaw's (2012) introduction parallels that of a traditional introduction. However, they present their overarching research questions through the use of a theatrical voice collage:

Jim: What drives us to write?

Joe: Where do we find our sources of inspiration?

Jim: What makes us want to record our inner thoughts for ourselves and for others?

Joe: What impact did the school system have on our ability to write and our attitudes towards writing?

Jim: What role did our informal writing play on our interest in writing?

Joe: Do different people answer such questions differently?

Jim: By addressing these questions through an exploration and interrogation of our writing histories, we may enable ourselves and other teachers to understand how their students approach writing. (p. 90)

Thus they maintain the integrity of the conversational style while introducing their research questions.

Sitter and Hall (2012) also use a conversational style, but rather than using questions as the frame, they "bracket in" (Norris, 2008) themselves. Sitter states:

> I'm wondering where we should begin? Prior to immersing ourselves in a conversation about boundaries, it may be helpful to start with discussing our different educational backgrounds. I've often found myself wondering how I ended up in social work but as a filmmaker?! (p. 245)

LeFevre and Sawyer (2012) employ a hybrid model. While their introduction contains the traditional content, it is embedded within the voice of each duoethnographer:

Rick: To investigate the difficulties encountered as well as the benefits gained when disclosing personal stories in research and teaching, we have chosen duoethnography as our research methodology...

Deidre: Yes, duoethnography is a conversation, and whenever we reveal ourselves, we become vulnerable. Such conversations can be both difficult to disclose and challenging. (p. 263)

All approaches are effective, but each style brings its own nuanced degree of the personal, or what Geertz (1974) would consider the "experience-near" (emic) or "experience-distant" (etic) stance. While duoethnographies are emic, the way that they are introduced can establish an etic tone. This has implications for writing decisions. Depending on the journal to which duoethnographic studies are submitted, the researchers may wish to adapt their style of introduction to better suit a particular readership.

Blending of Literature

While introductions may discuss background literature, duoethnographies, unlike many traditional studies, contain no discrete literature reviews. Rather, relevant literature is blended into the conversation as the writers deem appropriate. The literature, then, could be considered a third partner or discussant in a duoethnography, albeit one recalled by one of the duoethnographers. For example, Greenlaw stated: "I am reminded of Scholes' (1985) notion of 'textual power.' As he points out, we comprehend the writing of others by reading 'upon,' 'within,' and 'against' their texts" (Norris & Greenlaw, 2012, p. 93). In duoethnographies, two or more researchers bring in their knowledge of overlapping but sometimes disparate literature. Greenlaw's remark expanded Norris's understanding and he replied: "Ironically, by reading 'against' my text, you are working 'for' and 'with' me. Such is the power of collaborative writing" (p. 94). An exposure to the literature that one's writing companion knows is one of the pedagogical components of a duoethnographic partnership.

As indicated above, the literature can elucidate aspects of the research process, thereby embedding a discussion of the methodology within the conversation. It can also, however, provide insights on the topic under investigation. McClellan and Sader (2012) wrote, "As leadership educators we challenge ourselves to engage rather than deny or repress, differences that emerge at the dynamic intersections (Asher, 2007) of power, privilege, difference, and/or equity" (p. 138). They, and all duoethnographers, recognize the vital role that the literature plays in expanding one's awareness; however, the literature is not set apart as in many traditional texts.

The literature need not be from academic sources. Popular media, songs, and fiction are also parts of a duoethnographer's curriculum and can inform current analysis. Huckaby and Weinburgh's (2012) duoethnography is actually about the impact that the songs "Dixie" and "Lift Every Voice and Sing" had on their lives. Norris and Greenlaw (2012) discuss the influence of works by fiction writers Michael Ondaatje (1992) and Edgar Allan Poe (1962) in their curriculum of writing, and Sawyer and Liggett (2012) examine yearbooks. When writing a duoethnography, researchers are encouraged to look beyond academic texts and include other forms of media that have informed their curricula/ curreres and the topic.

Photographs and Artifacts

Photographs and artifacts not only assist in the recall of one's life experiences but are also useful displays to be embedded within a duoethnography. The first duoethnography (Norris & Sawyer,

2004) contained photographs taken throughout Sawyer's days growing up in Seattle. He discussed how the photographs themselves gave voice to the marginalized:

> In this photo, Steve is calling me a fruit. It was typical for us to appropriate the language of slurs and insults and make fun of them. Laughing at stereotypes of gay people was a way to process and in a sense contain them. We were not naïve, though, and we also played with images of entrapment and denied entitlement. (p. 147)

Shelton and McDermott (2012) include a picture of McDermott's sister putting lip gloss on her own daughter, making explicit the curriculum of beauty being passed down from mother to daughter. Sawyer and Liggett (2012) include a picture of Liggett in a bonnet in front of a fantasy "Snow White" house, demonstrating the engendered curriculum, and Norris and Greenlaw (2012) provide a newspaper clipping of Norris winning a public speaking contest in the summer between the seventh and eighth grades. The photographs add a texture that would not be present with the printed word alone.

Norris and Greenlaw (2012) also include a photograph of Norris's writing ideas book, an artifact that he once used to record thoughts, and both he and Greenlaw provide snippets of their creative writing, much of which was never published. Including these artifacts provide what Norris (1989) calls "contexture," or texture of the context that enhances the effectiveness of the text. In addition, through juxtaposition with the printed text, the pictures create an intertextual subtext as the artifacts generate new, nonverbal meanings for the readers.

Situating the Duoethnographers and Readers

As previously indicated, the conversational style of a duoethnography uniquely positions the reader by inhibiting an alliance with any one narrator. By having two or more people's stories being told in parallel, the reader is distanced from any one story. In a fashion similar to Brecht's alienation effect (1957), this style can enable critical thinking about the stories being told and the comments made about them. By being provided theses and antitheses with localized syntheses, readers can form their own syntheses.

The writing style in which two or more perspectives converse creates a third space (Bhabha, 1994) in which the readers can add their stories and perspectives while reading. Consequently the readers also become part of the construction process as the style in which the text is written evokes their own stories and insights. Duoethnographies are written to foster such a reading.

In the opening of a duoethnography on cross-cultural understanding (Aujla-Bhullar & Grain, 2012), Grain comments to Aujla-Bhullar, "Sonia, why don't you begin by telling us a little about who you are?" (p. 198). A lot is communicated by this phrase. First, Grain uses the second-person pronoun "you," making explicit that she is speaking with someone else. She also uses the first-person plural pronoun "us," indicating that she and others, the implied readers, will witness the response. With this one sentence, Grain informs the readers that they will be observing two people in conversation who are aware of their presence.

Breault, Hackler, and Bradley (2012) had the following insight early in their study: "As the project developed, Rick threw an element of necessary disequilibrium into the project.... We decided to add Rebecca into our conversation" (p. 122). Through the use of "we," they make explicit that there are others present, signaling that the readers will witness the stories and ideas of others. They continue with declarative prose, informing the readers about the study's methodology and content. There is little addressing of one another, and at times, they simply refer to each other in the third person, using the pronouns "he" or "she," rather than "I.". Their writing style makes the readers the people to whom the conversation is primarily directed, not the duoethnographic partners. In this study both the duoethnographers and readers are positioned differently from each other.

Norris and Greenlaw (2012) use a combination of structures that fluctuates between conversation and addressing an implied audience. After their introductory collage of questions, Norris directs this statement to the readers, "Due to its dialogic nature in examining life histories, we decided to use duoethnography" (p. 90). After this short direct address to the readers, Norris and Greenlaw launch into their conversation. Greenlaw opens with, "So, tell me, Joe, when did you first discover..." (p. 92). He later says, "It seems that your writing experiences during junior and senior high school years had some sense of audience" (p. 93), commenting directly on Norris's story. Throughout the study these researchers employ phrases like "Actually, Jim" (p. 94), "Like you, Joe" (p. 95) and "Unlike you, Jim" (p. 95). These maintain the conversational style throughout the chapter as well as acting as transitions or segues.

LeFevre and Sawyer (2012) also refer to each other by name throughout their study. Sawyer comments, "Deidre, your words take me back to Grade 3" (p. 265). LeFevre replies, "Rick, the importance of voice in writing is so critical, and yet you were in effect silenced through the 'curriculum' in your school at the time" (p. 266). In addition to their calling each other by name, they comment on each other's comments, situating each other as friendly critical colleagues.

While the name of the speaker is always provided in a duoethnography, its use within the conversation itself adds a personal tone to the writing. Readers know to whom the comments are directed. The intention of the shift to "we" is to inform the readers that they are being addressed. Explicit and subtle stylistic choices position the reader in relation to the text and duoethnographers in relation to one another. Such choices are at the researchers' disposal.

Finding/Creating a Collective Voice Through a Composite Narrator

While the use of the pronoun "we" provides a structure for a common voice, two duoethnographers take this further. Huckaby and Weinburgh (2012) have three distinct voices in their duoethnography:

1. Molly: I was 16 when you were born.
2. Fran: Sixteen years can make a difference...
3. Sojourners: Once again, the songs have given us a route to talk about difficult topics. (p. 167)

The addition of the Sojourners voice makes explicit that these words are spoken in unison, in harmony, with both parties in agreement. The Sojourners' words serve as short summaries or conclusions for each section of Huckaby and Weinburgh's duoethnography.

In the place of "Sojourners," LeFevre and Sawyer (2012) use the word "Both" to indicate sections of their duoethnography that reflect their common thoughts and general comments. In these sections, the pronoun shifts from first person singular (e.g., "I," "me," or "my") to first person plural (e.g., "we," "us," or "our").

Both: This duoethnography retraces some of our past experiences to generate new meanings about the nature of difficult or dangerous conversations. (p. 264)

Most duoethnographies imply a composite voice through the use of an unnamed narrator for text that is not preceded by one of the researchers' names. This traditional form of narration works quiet effectively. In the preceding two examples, however, the duoethnographers chose to name that voice. For Huckaby and Weinburgh, using the Sojourners voice serves as an implicit metaphor for using songs to find harmony in trying to understand and improve race relations. For LeFevre and Sawyer, "Both" acts as an overarching narrator. The choice of a general, impersonal narrator or a personified composite narrator offers duoethnographers another set of possibilities at their disposal.

Making the Showing and Telling Explicit Through Italics

The problematics of voice is a major issue that runs through all duoethnographies. Duoethnographers want to tell their stories and comment on them, but they recognize that both acts reflect points of view. They wish to both honor and disrupt their perspectives to maintain the dialogic quality of their writing. Two duoethnographies use italics as a means of making a distinction between explanation/telling and expression/showing.

Krammer and Mangiardi (2012) use a roman font when providing their stories (expression/showing) and italics when analyzing or commenting on them (explanation/telling).

Rose:...Because I successfully assimilated this logic, the person who bullied me the most was me.
Reconstructing these memories in the context of our duoethnography, I find myself thinking of my mother. (italics in original, p. 46)

Breault, Hackler, and Bradley (2012) reverse the structure: The stories are in italics and the commentary is in regular font.

Rick's Story

It is a weekend in the late spring of 1967, and a ten-year old boy sits on the porch of his home in suburban Chicago. Surrounding him today, instead of G.I. Joes or Matchbox cars, are his six-year-old sister and two of her girlfriends...

Since at least high school, with a few exceptions, Rick has enjoyed his time around the girls and women in his life more than time spent with his own sex. (p. 119)

While neither set of authors makes this structure explicit to their readers, in both cases it quickly comes apparent that the sections in italics provide different perspectives. The italics signal a shift from expression to explanation.

Transparency: Making Writing and Research Processes Explicit

Breault, Hackler, and Bradley (2012), unlike other duoethnographers, added hyperlinks to their text in the belief that doing so provides a greater degree of transparency. They write:

Even in the best of journal articles, the notion of transparency within conversational or narrative forms of research does not go much beyond a few extensive quotations from an even more extensive transcript and the author(s) [*sic*] insights and conclusions. Typically, the reader must trust that authors have honestly represented hours of dialogue in the selected quotations and that reams of field notes, participant reviews, and researcher notes indeed are reflected in a few pages of conclusions.

None of that is to imply that we need to be suspicious of all qualitative researchers. The lack of transparency, more often than not, is due to limitations of the publication process. Online journal formats offer the potential to free researchers from some traditional publication limits but only if the new format is used in nontraditional ways. In our project we experimented with presenting our work in a way that is as naturalistic and conversational as the original dialogue.

Inviting Others into the Conversation

> To accomplish our goal of creating a naturalistic and trans-
> parent presentation, we decided to use the hypertext capabil-
> ities of Web-based publication. All the work described below
> can currently be found at the *Trioethnography* link on http://
> breaultresearch.info. The materials included at that site are
> crucial to understanding our conversation and what we gained
> from it. This is especially true because in this chapter we have
> focused more on the process and method of our study than on
> the content related to gender construction. (pp. 125–126)

Breault, Hackler, and Bradley encourage readers to visit the web
site and contribute to the associated blog, thereby expanding the
conversation beyond the writers. The use of such technology can
add to the dialectic aim of duoethnographic studies by bringing
in the voices of readers. This is one possible future direction for
duoethnography.

While a method of this sort can increase dialogue, the desire
for such a level of transparency is debatable. Breault, Hackler, and
Bradley (2012) acknowledge that "Some fellow duoethnographers
will disagree with the need for so much attention to validity" (p.
128). As previously stated, all texts are constructed. Banks and
Banks (1998) claimed: "The opposite of fiction isn't truth but something like objectivity or actuality. Any genre or piece of writing
that claims to be objective, to represent the actual, is a writing that
denies its own existence" (p. 13). Duoethnographies are no more
nor less susceptible to the lack of rigor than any other research.
A paper trail is no guarantee of efficacy. Ultimately, it is up to the
readers to assess whether or not the writing tells and shows a story
in a manner that enables them to derive general insights that they
deem relevant. Transparency and rigor are embedded within the
conversation as the duoethnographers present and reconceptualize their stories in relation to the "other" (Dallery & Scott, 1989).
Dialectic conversations have their own internal rigor that becomes
apparent throughout the reading. Readers can ascertain whether
the degree of explanation and expression is effective in elucidating
particular aspects of the phenomenon under consideration and
whether enough of the methodology is made explicit throughout.
Hyperlinks are only one way of enabling the reader to do this.

Conclusion

In summary, when conducting a duoethnographic study and writing the final text, it is important for researchers to be aware of their epistemological, ontological, and axiological assumptions regardless of whether or not they make these explicit to the reader. These assumptions will and do guide both the form and content of what they write. Each writing choice will frame the study and situate the duoethnographers to their readers and to each other differently. Consequently, such choices are more than matters of style. They influence how the text will be experienced and its degree of credibility. Each duoethnographic partnership will debate and reach its own decisions regarding a variety of writing choices in addition to the storied and analytical content presented. This chapter has articulated a number of these choices to assist future duoethnographers.

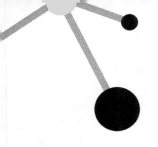

5

Temporal Insights, Placeholders, Stepping-stones, Milestones, and Epiphanies

WHILE TRADITIONAL RESEARCHERS most often end their writings by discussing conclusions, duoethnographers resist such definitive statements. There are few or no conclusions in duoethnographies. Life goes on. Things change. Meanings are contextually situated. While duoethnographies report transformations that evolved through conversation, these are not conclusive; rather they are situational in both time and place. But this does not mean that they have no value and are not generalizable. Barone (1990) claimed that readers conspire (or "breathe with," the etymological meaning) narrative texts, resonating with what they deem relevant. Generalizability does not rest with the researcher; rather, readers take what they read and generalize from particulars in one context, create a universal or parallel connection, and apply these generated meanings to their own contexts. By eavesdropping on the conversations of others readers can apply what they deem relevant to their past, present, and future lives.

Duoethnographers believe that their telling and coanalysis of their stories, in both their conversations and their writing, provides spaces for new meanings to emerge first of all for themselves and secondly for their readers. Their conversations are opportunities for insights as they juxtapose their respective understandings with the understanding of their research partner(s). By explicitly sharing these conversations, they invite readers to draw their own "insights," albeit in a different time and place. Rather than being declarative, claiming that their insights are universal, duoethnographers structure their conversations in a way that invites readers to draw their own conclusions, separate from the contextual ones of the duoethnographers. A "third space" in which dialogue can occur is created between each duoethnographer and between their finished text and their readers.

Routledge (1996), summarizing Bhabha's (1994) concept of the third space, claimed that "the third space is thus a place of invention and transformational encounters, a dynamic in-between space that is imbued with the traces, relays, ambivalences, ambiguities and contradictions, with the feelings and practices of both sites, to fashion something different, unexpected" (p. 406). Duoethnographers report these transformational encounters, create new ones for themselves, and, through their conversation, implicitly invite their readers to do the same with the text. The aim is to generate, through conversation, something unexpected. But rather than calling their reconceptualized understandings conclusions in the classical sense, duoethnographers prefer to view them as placeholders, stepping-stones, milestones, or temporal epiphanies, since changes can and will take place with each new experience.

In his discussion of currere, Pinar (1994) stated, "From another perspective, the method is self-conscious conceptualization of the temporal and from another, it is the viewing of what is that conceptualization through time" (p. 19). Duoethnography is a methodology of restorying and reconceptualization. Such temporal life changes are reported in duoethnographies, and through the research, duoethnographers expect not to reify previously held beliefs but to generate changes to old stories. Noting the temporal shifts of the duoethnographic process, Liggett wrote, "Fast-forwarding to the present, I began to make connections to my work and teaching, which focus on aspects and issues of racial and cultural identity and their impact on teaching and learning" (Sawyer & Liggett, 2012, p. 79).

Liggett's reconceptualization of the past reinformed her present understanding. These are the types of epiphanies experienced by duoethnographers. They not only report past understandings but, through revisiting them, create new ones. As Sawyer commented on his experience with memory in duoethnography:

> I have experienced my memories not as discrete linear images but, rather, as recursive thoughts. My earlier memories influenced my more recent memories and my more recent memories influenced my earlier ones—as I thought about them from my current moment in time. (LeFevre & Sawyer, 2012, p. 284)

This chapter summarizes the ways in which a variety of duoethnographers present their temporal epiphanies and insights about their topics, themselves, and the research process. At times, they report past events and beliefs as remembered, providing a timeline of progressions from a particular position to another. At other times, duoethnographers summarize their stories with plans for future actions. And at still other times, they gain insight about why it is necessary to maintain a current perspective or course of action. Predominantly, however, duoethnographies, as articulated above, consist of restorying past events with present insights. In that case all three kinds of insights are dialogically woven together, as their interplay elucidates the stories of both the researched and the research process.

Reporting the Past

The reporting of past events that influenced a duoethnographer's perspective on a phenomenon is a form of a temporal conclusion in that it situates a particular stance in a particular time. It provides readers with glimpses of what influenced beliefs and behaviors, and these reports act as historical milestones demarcating a curriculum/currere as it progresses. Norris (Norris & Sawyer, 2004) reports various stages in his learning and unlearning homophobia that evolved from his awareness of the phenomenon, through his questioning of imposed beliefs, and to realignment of his own perspective to advocacy for the rights of others of difference. He labels many of these stages "Becoming…" to indicate progression. He recalls:

> My first lesson in the curriculum of sexual orientation was one of homophobia. I learned that those who wore "green"

were fruits and that Thursday was fruit [day]. While the males had to wear white shirts and red ties as the school uniform requirement, I made sure that the pants my mother bought would never be the dreaded green. (p. 141)

During most on his time in elementary and secondary school, Norris believed in a social myth and behaved accordingly. Later, he rejected what he learned in the schoolyard. McDermott reports a similar experience with the social phenomenon of beauty:

The biggest contributor to my conception of beauty, in terms of people as beautiful, is my sister. She is 3 years older than I, and since my adolescence I remember her being the prettier one. I strived to find the right shoes, makeup, and hairstyle to meet her approval because I knew somehow that approval meant that I was "okay".... I also remember around the age of 13 getting reinforcement from men about my appearances. If I walked down the street and got honked at, I took that as a compliment. (Shelton & McDermott, 2012, p. 227)

At various stages of Norris's, McDermott's, and others' informal curricula/curreres, they reached particular conclusions about social mores and how they should act. While their entire texts make evident that these stepping-stones are not their present beliefs, the events at each stage provide a chronology of their social stances as experienced at specific times.

Like McDermott, Greenlaw concludes that others influenced his curriculum, this time within the school walls:

I was fortunate that year to have a very good English teacher named Doug Frame. Until then I had not connected my love of reading with school, but in Grade 11 my private interest in writing and my school life suddenly and irrevocably came together thanks to the kindness and skill of this English teacher. (Norris & Greenlaw, 2012, p. 197)

But while McDermott at times presents her thoughts in a purely informative fashion, implied within Greenlaw's text is a sense of gratitude that is a form of reflective conclusion. Subtle and explicit commentaries on a story can and do provide a form of meta-analysis or conclusion regarding the articulated experience.

Krammer makes this analytic awareness explicit by commenting directly on her school experience: "The hidden curriculum of my schooling taught me that success is predicated upon the defeat of others; and since life, by its very nature, denies us a perpetual winning streak, I came to fear those inevitable failures" (Krammer & Mangiardi, 2012, p. 49). She thereby provides her present conclusion about a past event. Krammer then considers a form of indoctrination, something that she only became aware of through her graduate studies:

> As an elementary and secondary student I was too naïve, and too well indoctrinated, to question school authority. It was unthinkable to blame my limited success on the mandated curriculum or the institutional structure.... Today I understand my academic "game-playing" as a compulsion born of the hidden curriculum of my schooling. (pp. 64–65)

Liggett also frames a past experience with a present editorial: "This provided fertile ground for shaking me to my very core as I stepped into the taxi queue outside the Nairobi airport. Thus began the slow process of deconstructing my own legacy of what it meant to be White" (Sawyer & Liggett, 2012, p. 74). While she was unaware of her curriculum at the time, her present recollection restories that event. Duoethnographers interlace such comments throughout their writing, extending or replacing old beliefs with new ones.

But not all positions are that decisive. Hall (Sitter & Hall, 2012) uncovers part of the root of his conflicting positions about professional boundaries within his family. He writes, "For example, to this day I can still recall my mother teaching me that discussing family differences should always occur within the privacy of the family, not in front of dinner guests or out in public. On one such occasion..." (p. 247). Hall never completely reaches a conclusion. Rather, his divisions among private, personal, and public are made problematic, allowing the reader to dwell within the challenge, the quest, the question.

In reporting past events duoethnographers may blend explanation and expression (Reason & Hawkins, 1988) by implicitly and explicitly commenting on the information that they provide. These commentaries are temporary microbeliefs, insights or conclusions that are contextual to time, place, and person. Ultimately it is up to readers to draw their own conclusions and implications for their own lives from the duoethnographers' research stories.

Realizations/Transformations/Reconceptualizations/ Resistance Through the Research Process

Duoethnographies, unlike traditional ethnographies, do not just report the given meanings that the participants hold. The research act is one of transformation. Duoethnographers recognize that being asked questions and listening to the stories of others may change their own perspectives. By retracing and reconceptualizing their respective curricula/curreres on a given phenomenon, duoethnographers blur the past and present as they develop new insights on previously held beliefs and either reconfirm or replace old beliefs These new insights are as close as duoethnographies come to conclusions.

McClellan and Sader (2012) together embrace the opportunity for personal reconceptualization:

As leadership educators we challenge ourselves to engage rather than deny or repress, differences that emerge at the dynamic intersections (Asher, 2007) of power, privilege, difference, and/or equity. The insights gained through this process will help us and our colleagues to create a safe space in the classroom through an examination of power and privilege in our own lives. (p. 139)

They recognize that they will change as a result of the research and that these changes, while sometimes challenging, will make new stepping-stones or placeholders in their thinking.

By reporting what they would consider both positive and negative aspects of their informal curricula of the phenomenon they are exploring, duoethnographers portray the social forces that foster and perpetuate particular values and beliefs. In Norris and Sawyer (2004), Norris retraced his curriculum of homophobia, making explicit the overt and convert ways that homophobia was propagated. Sawyer, on the other hand, retraced his counternarrative of resistance to heteronormativity in the temporal texts he and his friends produced. Lund, a social activist, did the same with his curriculum of racism. He reported:

Only a few rare times do I remember stopping to think about the impact my words might have on others. Once I was in a restaurant with a few friends, and I was telling distasteful "Paki" jokes and then realized there were a few people behind me who I believed were of Pakistani origin, and we laughed awkwardly. I felt self-conscious for a moment but never gave thought to the dehumanizing impact of this kind of humor. (Nabavi & Lund, 2012, p. 182)

Lund's writing partner, Nabavi, responded to him from a position of hope, demonstrating how previous behaviors can be understood and how that understanding can be used for future actions. She stated:

As we question the contexts and conditions that drove those early experiences for both of us in order to generate new meanings and insights, we also recognize that they inevitably have informed how we now engage with issues of social and political justice. We share a common bond in our commitment to transform, reform, and reinscribe our interpretations of pejorative acts we were each subject to and perpetuators of. (p. 183)

For Norris, Lund, and other duoethnographers, the act of reconceptualizing and restorying one's life, while not therapy, can be therapeutic. Lund wrote:

I remain deeply ashamed of the person I was, and I wonder if there is an ongoing effort to assuage my lingering (White) guilt embedded in the work I now do for a living. In many ways I feel damaged by this kind of behavior—not to minimize the subjects of racism, but I know that there are many victims of oppression, *including* the apparent perpetrators. (Nabavi & Lund, 2012, p. 183)

Duoethnographies provide cathartic moments for duoethnographers as they take what they perceive to have been previous negative behaviors and beliefs and transform them by exposing them. The reporting of the stages of one's curriculum allows readers to witness a phenomenon as lived, warts and all. Aujla-Bhullar and Grain (2012) discuss this transformative experience:

Kari: Yeah, and it's funny; I feel uncomfortable saying these things because it seems like such a sham, trying to reconcile these critical thoughts with my White identity.

Sonia: It's healthy, though, to have those thoughts and share them in a context that is safe and trusting. (p. 210)

In addition to providing reports about their actions, Aujla-Bhullar and Grain also give accounts of their lack of action in encounters with peers who exhibited racist behaviors. Grain comments, "it's interesting that your experience of regret comes from failing to defend your own culture whereas my experience of regret comes from failing to defend someone else's" (p. 206). Such comments act like temporal conclusions as the duoethnographers reinscribe former experiences with present analyses.

Furthermore, via counternarratives and personal knowledge, duoethnographers critique more hegemonic meanings, narratives, and institutional structures. For example, McClellan and Sader (2012), in their duoethnography on power and privilege, discuss academic views of who should teach diversity courses in higher education settings:

> We understand that Jennifer can't fully understand what it is like to be Black, but assuming that Patrice is somehow an "expert" in diversity just because of her standpoint as a Black woman is to essentialize her unfairly and discount her scholarship in this area. As Collins (2000) rightly pointed out, "despite the common challenges confronting African-American women as a group, individual Black women neither have identical experiences nor interpret experiences in a similar fashion" (p. 27). Patrice's expertise in social justice comes from her study of the issues and theory, not just from experiences. The point of this class is to initiate conversations around privilege and power and their relationship to race, gender, sexual orientation, religion, and other sites of prejudice. We couldn't give the sense that we agreed that the problem of confronting racism belongs to Blacks, and that Whites have no responsibility to start and participate in these conversations. Neither of us wanted to be complicit in a system that models passive acceptance of privilege and oppression by agent groups. (p. 145)

Clearly, instead of changing their perspectives, they both further resolved to reveal and resist a flawed system.

Duoethnographers report such advances in their thinking in a number of ways. They gain new insights about the phenomenon they are investigating, about themselves, and about the methodology. Rather than appearing as discrete units, usually all three are blended with the research conversation. However, we present them separately below to indicate the variety of conclusions and insights that can be found within duoethnographic texts.

Types of Insights

Insights on the Phenomenon

As previously discussed, duoethnographers are the sites, not the topic of the research. While insights about self are to be expected, the focus of the research is the understanding of the phenomenon through self. Mangiardi, in an exploration of the hidden curriculum, first changes one of her beliefs and then uses her new understanding in a critique of society:

> I recognize that the duoethnographic process has helped transform my understanding of "success." Success, of course, is a fundamental concept in education, implicit in any discussion of students and schooling. Yet, despite its prevalence and significance, educators and students rarely discuss its meanings and their ramifications. As a result, teachers send students contradictory messages about success. (Krammer & Mangiardi, 2012, p. 62)

Mangiardi uses the self to assist in the reexamination and critique of a common educational phenomenon, bringing a fresh perspective on an entrenched traditional practice. Readers witness a progression of thought as she comes to a new, more fluid stance. Such insights on phenomena emerge throughout a duoethnography, flowing organically through the conversation.

While Mangiardi uses personal knowledge to inform public knowledge, Weinburgh reverses the process (Huckaby & Weinburgh, 2012). The research expands her own beliefs as she becomes aware of different perspectives on the song "Dixie":

> The phrase "live and die for Dixie" meant something different for me. But now, knowing who wrote it, when it was written, how one dressed in blackface to sing it, the vernacular that made it what it is: These things make the song different for me; it is about not home but servitude. (p. 165)

Insights in duoethnographies surface recursively from the interplay of the emic (the self) and the etic (the other, including the literature and research partners).

While some duoethnographies question and displace previously held perspectives on a phenomenon, others extend previous meanings. Hall expanded his understanding of the range of professional boundaries as they are defined by participants and external perspectives:

> One thing that struck me when having this discussion was the notion of boundaries as some sort of fixed or defined entity. As our discussion progressed, it really reaffirmed for me the idea that boundaries truly are unique to the time, place, and topic of discussions and specific relationships. At first glance this might be taken to mean boundaries are merely situational or conditional, but I think that this is not the case at all; it is much more than this. Boundaries can and do change, but this seems much more about that space created between participants than some other outside force. (Sitter & Hall, 2012, p. 259)

LeFevre, in contrast to Mangiardi and Weinburgh, problematized the process of personal insight, articulating that gaining multiple perspectives on a phenomenon can be at times risky:

> The Internet can magnify a sense of danger in terms of the public, permeable, and permanency issues it layers on conversations. However, at the same time, the opportunities for communication these new technologies offer are also powerful and liberating. So it's important to hold on to the big picture, including the positive aspects of ICT [informations and communications technologies] for communication. (LeFevre & Sawyer, 2012, p. 279)

In this case, her stance is neither displaced nor extended. Rather, she embraces ambiguity. In a postmodern era, conclusions themselves become problematic.

As the above examples show, the discovery or creation of insights through duoethnographic research follows no preset formula. Surprises are common as sifting the material of the conversational data trail reveals gems that advance personal and public knowledge of the phenomenon under consideration, whether by displacing or extending them.

Insights on Self

One cannot completely separate the knower from the known, the person from the meaning of the phenomenon, as beliefs are intricately linked to self. While this fact is implicit in many research genres, it becomes apparent in narrative works such as duoethnography. Shelton directly links her understanding of beauty to her concept of self:

> Doing this research has made me think about my construction of self differently than what I have done in the past. I have examined my academic self. I have asked myself: Why I am the only PhD in my family? Why did I change my linguistic identity? Why did I disassociate myself from my Upstate New York identity for so many years? But this duoethnography also made me realize that my conception of beauty is a different conception of self. It is completely different from my academic or linguistic self. Yet, I now see it as part of my drive to be who I am academically and linguistically. I think since I have always, deep down, felt a little less than "pretty," I have worked very hard to find a power in my brain, in my language (especially in writing), and in my career. (McDermott & Shelton, 2012, p. 238)

Sawyer also reports a change in self-perception, one that contradicted his previous belief:

> I would like to think that I have published dangerous conversations throughout my professional life, but as I examine my publication history, I see that this really isn't true.... I now ask myself, What stands out for me as something I consider a controversial piece of writing? I have written a number of pieces in the past where I have championed the rights of others, but rarely did I write anything that created a risk for myself. There is one exception. Joe Norris and I wrote the first duoethnography that explored aspects of our lives where our identities had crossed into a range of sexual identification territories. (LeFevre & Sawyer, 2012, p. 271)

Duoethnographies can and do hold up a mirror to researchers, challenging them to delve deeply into what they believe. At times, self-deception may become evident and overcome as the process elicits new epiphanies.

While Shelton and Sawyer articulate an increased awareness of self, Norris (Norris & Greenlaw, 2012) reports a change in attitude and behavior: "Actually, Jim, since our first draft of this, I have come to enjoy editing more. I have my first solo book published..., and I relish the craft of phrasing. So editing now is far less painful" (p. 100). For him the insights gained through duoethnography, coupled with other life experiences, not only changed his attitude toward certain behaviors but also changed the behaviors themselves. Had Sawyer commented on plans for future writing with a riskier stance, his insight too would have moved from a change in awareness to a change in behavior.

Lund implicitly undergoes such a shift in his critique of and challenge to self:

> But now, even looking at my work with those students, I was so often lauded for "leading" students with this project that I rarely found occasion to trouble and question my own latent assumptions and unearned privileges. Living my life as a "normal" Canadian, I was never directly challenged to understand the implications of my White, straight, able-bodied male identity. Even being in this field of study, I am rarely invited to implicate my own identity in my work. As a cultural insider, I never have my *Canadianness* questioned, and my sense of belonging, citizenship, and entitlement seems quietly affirmed. (Nabavi & Lund, 2012, p. 187)

By claiming that his "Canadianness" in not questioned, Lund in fact questions it. The articulation of changes in awareness can implicitly indicate a change in behavior because one makes explicit to others an otherwise unreported position. Insights about self become part of the duoethnographic process as they simultaneously shed light on the phenomenon under investigation.

Insights on the Methodology

Just as one cannot completely separate the known from the knower, one cannot completely separate form from content, the message from the medium (McLuhan, 1977), or the methodology from the data. Data is [media]ted by methodology. Bradley comments on how the methodology of duoethnography assisted her in a change of perspective: "This process was insightful for me because it challenged my somewhat negative perceptions of male elementary school teachers.... Through this process, I have begun to see an alternate role for male elementary school teachers" (Breault, Hackler, & Bradley, 2012, p. 133).

That the methodology directly influenced Bradley's change of perspective makes evident that it is partially about personal transformations. As McClellan reported:

> I walk away from this duoethnography having experienced a deeper awareness of myself as a leadership educator/social justice educator. There is an old adage that says, "Understanding takes time." Having heard that for most of my childhood, it seemed like a misnomer. I knew the significant meaning of the phrase in an intellectual way. However, through this duoethnography I've experienced the tangibility of its meaning. To understand myself in the reflection of Jennifer was a vulnerable feeling. At times I wanted to quit for fear of the unknown and many questions that followed, such as, What would our relationship be like after this process as colleagues? What if I share too much? What if she doesn't share at all? And when will the process of discovery end? But I was able to learn through the dissonance and discomfort, as Griffin and Ouellett (2007) suggest, and began to experience comfort in the relationship that I was building with Jennifer. (McClellan & Sader, 2012, p. 152)

Not only does McClellan articulate her increased understanding of the phenomenon, she also makes explicit her internal personal struggles with the methodology. Since duoethnographers are the sites of their research, there is a vulnerability regarding the unanticipated changes in awareness and behavior that might emerge through the methodology—the vulnerability of making those changes known to others and of one's relationship with one's fellow duoethnographer. Comments like McClellan's are evidence of rigor that demonstrates that the methodology goes beyond navel-gazing to introspection and reconceptualization.

Aujla-Bhullar believes that the methodology has a cathartic dimension that brings focus to her work:

> I feel comfortable sharing these narratives and tracing my learning to certain moments and specific types of curriculum.

> I hadn't really thought of travel as a type of curriculum or my experiences with bullying as a type of curriculum until now. And through these very different types of curriculum, I've formed my idea of social justice and identity and racism, among other things. It's almost as if talking about it with you has pulled those past experiences into the present and repositioned them in my adult frame of reference. Then to juxtapose all of this with your narratives of learning… it's kind of enlightening. (ellipsis in original, Aujla-Bhullar & Grain, 2012, pp. 220–221)

Clearly, for her, the methodology is transtemporal, enabling multiple perspectives on a single event by a single person. Such insights on the methodology are often fully integrated with discussion of the phenomenon under investigation and not compartmentalized into a remarks section solely about the methodology.

While transformative and cathartic, duoethnography is no panacea. The insights discovered only offer potential future directions. Huckaby and Weinburgh (2012) clearly recognize that the methodology does not offer stable conclusions:

> As we try to answer all these questions posed and even those left unstated, we find they only lead to more questions, layered like an onion. We've been trying to figure out a metaphor for these conversations because we've been under the illusion that at some point this might get to something. And we've gotten to many things, some of which are in this text. Could it be like the mathematical one half, where there will always be a half that we cannot traverse, Zeno's paradox. It's just always half, half, half; no matter how close we get or how many steps we take, if we take half steps toward each other, we will always be taking half steps; if we do it by halves, we will never be able to walk into the other. But, we can approach the other and get closer than we might have imagined possible. (p. 175)

For Huckaby and Weinburgh and other duoethnographers, the only conclusion is the willingness to engage with the other. Duoethnographic research is about creating a meeting place for in-depth discussion. From such events, the participants/pilgrims leave renewed by the stories of others (Kopp, 1972). By focusing and editing their conversations and sharing them with others, duoethnographers hope to continue the conversation, with the belief that readers or audience members will create their own syntheses or temporal insights that are contextual to their own times and places.

In their duoethnography texts, duoethnographers do not represent their own and others' meanings. To do so would suggest that the meanings lie outside themselves. Rather, they phenomenologically transact with the text, creating their meanings of "reality" in a dialogic process. This is a key contribution of duoethnography to the dilemma of representation in qualitative research. Instead of building trustworthiness through triangulation of data collection, they show it through their engagement and praxis within the texts they create.

Action and Future Plans

Duoethnographies do not remain "talks among leisured people" (Tolstoy, 1966). Duoethnographers report how they have implemented or plan to implement their findings as actions. In the first duoethnography (Norris & Sawyer, 2004), Norris reports how he spoke against homophobia at a parent–teacher meeting discussing sex education, and in another duoethnography, Sawyer questions the "overall framing of power and privilege in language" as it relates to grading papers (Sawyer & Liggett, 2012, p. 84). Duoethnography is not just about reporting one's life but also about taking action based on the temporary conclusions or ambiguous insights gleaned.

For example, Hackler states, "As I listened to the stories, reflections, and questions of the two of you, I began to understand and validate the relevance of events in my own professional and personal journey" (Breault, Hackler, & Bradley, 2012, p. 132). But he does not remain at the reporting stage. As a new principal, he constructs opportunities for his staff to explore, question, and challenge their work experience and structures of education (p. 133). For Hackler, this new awareness came with a new set of responsibilities.

Liggett uses her increased awareness to direct her teaching. This enables her to be more vigilant in detecting assumptions:

> As a teacher educator I try to create shifts in perspective through course readings, media, and activities geared toward

evoking transformative experiences, shifts that uncover culturally nuanced values and expectations that inform teaching—in a sense, a hidden cultural curriculum that can disadvantage students who are not members. Even as I work toward deconstructing dominant cultural assumptions in my classes, I catch myself making assumptions about my students, colleagues, friends, family, and the various people I encounter throughout my day. This process of deconstruction is slippery; it takes constant vigilance. (Sawyer & Liggett, 2012, p. 85)

Such comments are conclusions inasmuch as they indicate a determination to act. They are, however, not didactic or prescriptive. They are guidelines that create a sensitivity toward the phenomenon from which action can arise. Krammer and Mangiardi (2012) clearly articulate such a stance:

Most important, our duoethnographic inquiry has taught us that it is never enough to expose a hidden curriculum; we must also understand *how* that curriculum accomplishes its ideological work. We must go beyond exposure and aim for comprehension. When we identify what our hidden curricula have taught us, we take only the first step. We take a second step when we recognize the institutional selves formed by those lessons. But it is only when we comprehend *how* the lessons have been presented and the ideologies transmitted that we are able to apply our research and are empowered to instigate change. (p. 68)

Conclusion (or Not)

This chapter reconceptualizes the term "conclusions" in both definition and format. First, conclusions in duoethnographies are not conclusive; they are placeholders or stepping-stones until another experience or reflection instigates a change in perspective and action. Duoethnographers report how they have experienced such milestones or epiphanies throughout their lives and foreshadow future ones. Second, through conversations, duoethnographers learn from each other, and the duoethnography is in its own right a milestone that marks a significant event of reflection and reconceptualization. Third, while summaries may exist in a duoethnography, because insights are predominantly reported in chronological order, readers cannot assume they will find a conclusions section at the end of a study. Rather, they must read the researchers' stories to contextualize the reported insights. Duoethnographers blend "explanation" and "expression" (Reason and Hawkins, 1988) as they tell and show their stories about a particular phenomenon.

6

References and General Readings

References

Aoki, T. (2005). Toward curriculum inquiry in a new key. In W. F. Pinar & R. Irwin (Eds.), *Curriculum in a new key* (pp. 89–110). Mahwah, NJ: Lawrence Erlbaum.

Asher, N. (2007). Made in the (multicultural) U.S.A.: Unpacking tensions of race, culture, gender, and sexuality in education. *Educational Researcher, 36*(2), 65–74.

Aujla-Bhullar, S., & Grain, K. (2012). Mirror imaging diversity experiences: A juxtaposition of identities in cross-cultural initiatives. In J. Norris, R. D. Sawyer, & D. Lund (Eds.), *Duoethnography: Dialogic methods for social, health, and educational research* (pp. 199–222). Walnut Creek, CA: Left Coast Press.

Ayers, W., & Miller, J. (1998). *A light in dark times: Maxine Greene and the unfinished conversation.* New York, NY: Teachers College Press.

Bakhtin, M. M. (1981). *The dialogic imagination.* Austin, TX: University of Texas Press.

Banks, A., & Banks, S. (1998). *Fiction and social research: By ice or fire.* Walnut Creek, CA: AltaMira Press.

Barone, T. E. (1990). Using the narrative text as an occasion for conspiracy. In E. W. Eisner & A. Peshkin (Eds.), *Qualitative inquiry in education* (pp. 305–326). New York, NY: Teachers College Press.

Bateson, M. C. (1994). *Peripheral visions: Learning along the way.* New York, NY: HarperCollins.

Becker, C. (Ed.) (1994). *The subversive imagination: Artists, society, and social responsibility*. New York, NY: Routledge.

Behar-Horenstein, L. S., & Morgan, R. (1995). Narrative research, teaching, and teacher thinking: Perspectives and possibilities. *Peabody Journal of Education, 70*(2), 139–161.

Bhabha, H. (1994). *The location of culture*. New York, NY: Routledge.

Bowers, C. A. (1997). *The culture of denial*. Albany, NY: State University of New York Press.

Bradbeer, J. (1998). *Imagining curriculum: Practical intelligence in teaching*. New York, NY: Teachers College Press.

Breault, R., Hackler, R., & Bradley, R. (2010). *Tri-ethnography: The construction of male identity in two elementary school teachers*. Retrieved January 5, 2011, from http://breaultresearch.info/transcript-hour-2.html

Breault, R., Hackler, R., & Bradley, R. (2012). Seeking rigor in the search for identity: A trioethnography. In J. Norris, R. D. Sawyer & D. Lund (Eds.), *Duoethnography: Dialogic methods for social, health, and educational research*. Walnut Creek: Left Coast Press.

Brecht, B. (1957). *Brecht on theatre: The development of an aesthetic* (J. Willett, Trans.). New York, NY: Hill and Wang.

Ceglowski, D. (in press). Duoethnography with children. *Ethnography and Education*.

Chang, H. (2008). *Autoethnography as method*. Walnut Creek, CA: Left Coast Press.

Charmaz, K. (2009). Shifting the grounds: Constructivist grounded theory methods. In J. M. Morse, P. N. Stern, J. Corbin, B. Bowers, K. Charmaz, & A. E. Clark (Eds.), *Developing grounded theory: The second generation* (pp. 127–192). Walnut Creek, CA: Left Coast Press.

Christians, C. G. (2005). Ethics and politics in qualitative research. In N. K. Denzin & Y. S. Lincoln (Eds.), *The SAGE handbook of qualitative research* (3rd ed., pp. 139–164). Thousand Oaks, CA: Sage.

Clandinin, J. (2011, January). *Keynote Address*. Paper presented at the Narrative, Arts-based, and "Post" Approaches to Social Research Conference, Tempe, AZ.

Clandinin, J. D., & Connelly, F. M. (1992). Teacher as curriculum maker. In P. W. Jackson (Ed.). *Handbook of research in curriculum* (pp. 402–435). New York, NY: Macmillan.

Clandinin, J. D., & Connelly, F. M. (2000). *Narrative inquiry: Experience and story in qualitative research*. San Francisco, CA: Jossey-Bass.

Coles, R. (1989). *The call of stories: Teaching and the moral imagination*. Boston, MA: Houghton Mifflin.

Connelly, F. M., & Clandinin, D. J. (1990). Stories of experience and narrative inquiry. *Educational Researcher, 19*(5), 2–14.

Creswell, J. W., & Miller, D. (2000). Determining validity in qualitative inquiry. *Theory into Practice, 39*(3), 124–129.

Cunningham, I. (1988). Interactive holistic research: Researching self managed learning. In P. Reason (Ed.), *Human inquiry in action: Developments in new paradigm research* (pp. 163–181). Newbury Park, CA: Sage.

Czarniawska, B. (1997). *Narrating the organization: Dramas of institutional identity*. Chicago, IL: University of Chicago Press.

Dallery, A., & Scott, C. (1989). *The question of the Other*. Albany, NY: State University of New York Press.

Denzin, N. K. (1997). *Interpretive ethnography: Ethnographic practices for the 21st century*. Thousand Oaks, CA: Sage.

Denzin, N. K. (2003). *Performance ethnography: Critical pedagogy and the politics of culture*. Thousand Oaks, CA: Sage.

Derrida, J. (1973). *Speech and phenomena: And other essays on Husserl's theory of signs* (D. B. Allison, Trans.). Evanston, IL, Northwestern University Press.

Derrida, J. (1985). *The ear of the other*. Lincoln, NE: University of Nebraska Press.

Diversi, M., & Moreira, C. (2009). *Betweener talk: Decolonizing knowledge production, pedagogy, & praxis*. Walnut Creek, CA: Left Coast Press.

Donaldson, S. (1999). *Hemingway vs. Fitzgerald: The rise and fall of a literary friendship*. Woodstock, NY: Overlook Press.

Ebert, R. (1999). *My Dinner with Andre* (1981) [Review of the motion picture]. Retrieved January 3, 2011, from http://rogerebert.suntimes.com/apps/pbcs.dll/article?AID=/19990613/REVIEWS08/906130301/1023

Eisner, E. W. (1991). *The enlightened eye: Qualitative inquiry and the enhancement of educational practice*. New York, NY: Macmillan.

Ellis, C. (2004). *The ethnographic I: A methodological novel about autoethnography*. Walnut Creek, CA: AltaMira Press.

Ellis, C., & Bochner, A. P. (2000). Autoethnography, personal narrative, reflexivity: Researcher as subject. In N. K. Denzin & Y. S. Lincoln (Eds.), *The SAGE andbook of qualitative research* (2nd ed., pp. 733–768). Thousand Oaks, CA: Sage.

Fisher, G., & Sedgwick, E. K. (1996). *Gary in your pocket. Stories and notebooks of Gary Fisher*. Durham, NC: Duke University Press.

Foucault, M. (1980). *Power/knowledge: Selected interviews and other writings*. New York, NY: Pantheon.

Foucault, M. (2003). *Society must be defended : Lectures at the Collège de France, 1975–76* (D. Macey, Trans.). New York, NY: Picador.

Freire, P. (1970). *Pedagogy of the oppressed*. New York, NY: Seabury Press.

Freire, P. (1973). *Pedagogy of the oppressed*. New York, NY: Seabury Press.

Gadamer, H. G. (1975). *Truth and method*. London, UK: Sheed and Ward.

Geertz, C. (1973). Thick description: Toward an interpretive theory of culture. In C. Geertz (Ed.), *The interpretation of cultures*. New York, NY: Basic Books.

Geertz, C. (1974). From the native's point of view: On the nature of anthropological understanding. In P. Rabinow & W. Sullivan (Eds.), *Interpretive social sciences* (pp. 221–237). Berkeley, CA: University of California Press.

Greene, M. (1991). Teaching: The question of personal reality. In A. Lieberman & L. Miller (Eds.), *Staff development for education in the 90s: New demands, new realities, new perspectives*. New York, NY: Teachers College Press.

Greene, M. (2001) Foreward. In G. Dimitriadis & C. McCarthy, C. (Eds.), *Reading and teaching the postcolonial: From Baldwin to Basquiat and beyond* (pp. vii–viii). New York, NY: Teachers College Press.

Greenlaw, J., & Norris, J. (2012). Responding to the Muses: Two interwoven autoethnographic journeys on writing. In J. Norris, R. D. Sawyer, & D. E. Lund (Eds.), *Duoethnography: Promoting personal and societal change in dialogic self-study* (pp. 89–113). Walnut Creek, CA: Left Coast Press.

Greenwood, D. A. (2009). Place, survivance, and White remembrance: A decolonizing challenge to rural education in mobile modernity. *Journal of Research in Rural Education, 24*(10). Retrieved November 27, 2010, from http://jrre.psu.edu/articles/24-10.pdf

Griffin, P. & Ouellett, M. L. (2007). Facilitating social justice education courses. In M. Adams, L. A. Bell, & P. Griffin (Eds.), *Teaching for diversity and social justice* (2nd ed., pp. 89–113). New York, NY: Routledge.

Grumet, M. R. (1988). *Bitter milk: Women and teaching.* Amherst, MA: University of Massachusetts Press.

Guba, E. G., & Lincoln, Y. S. (1994). Competing paradigms in qualitative research. In N. K. Denzin & Y. S. Lincoln (Eds.), *The SAGE handbook of qualitative research* (pp. 105–117). Thousand Oaks, CA: Sage.

Haven, K. (2007). *Story proof: The science behind the startling power of story.* Westport, CT: Libraries Unlimited.

Hellbeck, J. (2010, December). The maximalist: On Vasily Grossman. *The Nation, 291*(25), 25–30.

Holland, D., Lachicotte, W., Jr., Skinner, D., & Cain, C. (1998). *Identity and agency in cultural worlds.* Cambridge, MA: Harvard University Press.

Holquist, M. (1981). Introduction. In M. M. Bakhtin, *The dialogic imagination* (pp. xv–xxxiii). Austin, TX: University of Texas Press.

Huckaby, M. F., & Weinburgh, M. (2012). Alleyways and pathways: Our avenues through patriotic songs. In J. Norris, R. D. Sawyer, & D. Lund (Eds.), *Duoethnography: Dialogic methods for social, health, and educational research* (pp. 157–176). Walnut Creek, CA: Left Coast Press.

Iannacci, L. (2007). Critical narrative research (CNR): Conceptualizing and furthering the validity of an emerging methodology. *Vitae Scholasticae, 24*, 55–76.

Janesick, V. J. (2000). The choreography of qualitative research design: Minuets, improvisation, and crystallization. In N. K. Denzin & Y. S. Lincoln (eds.), *The SAGE handbook of qualitative research* (2nd ed., pp. 379–399). Thousand Oaks, CA: Sage.

Kim, J. (2006). For whom the bell tolls: Conflicting voices inside an alternative high school. *International Journal of High School and the Arts, 7*(6), 1–9.

Kincheloe, J. L. (2001). *Getting beyond the facts: Teaching social studies/social sciences in the twenty-first century.* New York, NY: Peter Lang.

Kopp, S. (1972). *If you meet the Buddha on the road, kill him.* New York: Bantam.

Koro-Ljungber, M., Yendol-Hoppey, D., Smith, J. J., & Hayes, S. G. (2009). (E)pistemological awareness, instantiation of methods, and uninformed methodological ambiguity in qualitative research projects. *Educational Researcher, 38*(9), 687–699.

Krammer, D., & Mangiardi, R. (2012). A duoethnographic exploration of what schools teach us about schooling. In J. Norris, R. D. Sawyer, & D. Lund (Eds.), *Duoethnography: Dialogic methods for social, health, and educational research* (pp. 41–70). Walnut Creek, CA: Left Coast Press.

Kunstler, J. H. (1993). *The geography of nowhere: The rise and decline of America's man-made landscape.* New York, NY: Simon & Schuster.

Kurosawa, A. (Director). (1950). *Rashōmon* [Motion picture]. Japan: Daiei Motion Picture Company.

Langer, S. (1942). *Philosophy in a new key: A study in the symbolism of reason, rite and art.* Cambridge, MA: Harvard University Press.

Lather, P. (1986). Issues of validity in openly ideological research: Between a rock and a soft place. *Interchange, 17*(4), 63–84.

Latour, B. (1993). *We have never been modern.* Cambridge, MA: Harvard University Press.

Leavy, P. (2009). *Method meets art.* New York, NY: Guilford Press.

LeFevre, D. M. L., & Sawyer, R. D. (2012). Dangerous conversations: Understanding the space between silence and communication. In J. Norris, R. D. Sawyer, & D. Lund (Eds.), *Duoethnography: Dialogic methods for social, health, and educational research* (pp. 261–287). Walnut Creek, CA: Left Coast Press.

Lévinas, E. (1984). Emmanuel Lévinas. In R. Kearney (Ed.), *Dialogues with contemporary Continental thinkers* (pp. 47–70). Manchester, UK: Manchester University Press.

Lewin, K. (1948). *Resolving social conflicts.* New York, NY: Harper and Brothers.

Liggett, T., & Sawyer, R. D. (2009, April). *Post-colonial education: Duoethnography as an act of decolonization.* Paper presented at the Annual Meeting of the American Educational Research Association, San Diego, CA.

Lund, D. E., & Nabavi, M. (2008). A Duo-ethnographic conversation on social justice activism: Exploring issues of identity, racism, and activism with young people. *Multicultural Education, 15*(4), 27–32.

MacIntyre, A. (1981). *After virtue: A study in moral theory.* Notre Dame, IN: University of Notre Dame Press.

Malle, L. (Director). (1981). *My Dinner with Andre* [Motion picture]. United States: Saga Productions.

Marable, M. (2007). *Living black history: How reimagining the African-American past can remake America's future.* New York, NY: Basic Civitas Books.

Maruyama, M. (2004). Peripheral vision: Polyocular vision or subunderstanding? *Organization Studies, 25*(3), 467–480.

McClellan, P., & Sader, J. (2012). Power and privilege. In J. Norris, R. D. Sawyer, & D. Lund (Eds.), *Duoethnography: Dialogic methods for social, health, and educational research* (pp. 137–156). Walnut Creek, CA: Left Coast Press.

McDermott, M., & Shelton, N. R. (2008, April). *A curriculum of beauty.* Paper presented at the Annual Meeting of the American Educational Research Association, New York, NY.

McDermott, M., & Shelton, N. R. (2012). A curriculum of beauty. In J. Norris, R. D. Sawyer, & D. E. Lund (Eds.), *Duoethnography: Promoting personal and societal change in dialogic self-study* (pp. 223–242). Walnut Creek, CA: Left Coast Press.

McLaren, P. (1989). *Life in schools: An introduction to critical pedagogy in the foundations of education.* New York, NY: Longman.

McLuhan, M. (1977). *The medium is the massage.* New York, NY: Random House.

McMillan, J. H., & Schumacher, S. (1989). *Research in education: A conceptual introduction.* Glenview, IL: Scott, Foresman.

Merleau-Ponty, M. (1962). *Phenomenology of perception* (C. Smith, Trans.). New York, NY: Humanities Press.

Miller, J. (2011, January). *Autobiography on the move: Poststructuralist perspectives on (im)possible narrative representatives of collaboration.* Paper presented at the Narrative, Arts-Based, and Post Approaches to Social Research Conference, Tempe, AZ.

Mills, G. E. (2007). *Action research: A guide for the teacher researcher.* Upper Saddle River, NJ: Pearson Education.

Morris, D. (2002). Narrative, ethics, and pain: Thinking with stories. In R. Charon & M. Montello (Eds.), *Stories matter: The role of narrative in medical ethics.* New York, NY: Routledge.

Muncey, T. (2010). *Creating autoethnographies.* Thousand Oaks, CA: Sage.

Nabavi, M., & Lund, D. E. (2010, April). *Activism in Schools.* Paper presented at the Annual Meeting of the American Educational Research Association, Denver, CO.

Nabavi, M., & Lund, D. E. (2012). Tensions and contradictions of living in a multicultural nation in an era of bounded identities. In J. Norris, R. D. Sawyer, & D. Lund (Eds.), *Duoethnography: Dialogic methods for social, health, and educational research* (pp. 177–197). Walnut Creek, CA: Left Coast Press.

Neilsen, A. (1999). *Daily meaning: Counternarratives of teachers' work.* Mill Bay, British Columbia, Canada: Bendall Books.

Noe, A. (2000). Experience and experiment in art. *Journal of Consciousness Studies, 7*(8–9), 123–135.

Norris, J. (1989). *Some authorities as co-authors in a collective creation production* (Unpublished doctoral dissertation). University of Alberta, Edmonton, Alberta, Canada.

Norris, J. (2008). Duoethnography. In L. M. Given (Ed.), *The Sage encyclopedia of qualitative research methods* (Vol. 1, pp. 233–236). Los Angeles, CA: Sage.

Norris, J. (2009). Playbuilding as qualitative research: A participatory arts-based approach. Walnut Creek: Left Coast Press.

Norris, J., and Greenlaw, J. (2005, October). *Response to the muses: Two interwoven autoethnographic journeys of writing.* Paper presented at the Curriculum and Pedagogy Conference, Oxford, OH.

Norris, J., & Greenlaw, J. (2012). Responding to our muses: A duoethnography on becoming writers In J. Norris, R. D. Sawyer, & D. Lund (Eds.), *Duoethnography: Dialogic methods for social, health, and educational research* (pp. 89–113). Walnut Creek, CA: Left Coast Press.

Norris, J., & Sawyer, R.D. (2004). Hidden and null curriculums of sexual orientation: A dialogue on the curreres of the absent presence and the present absence. In L. Coia, N. J. Brooks, S. J. Mayer, P. Pritchard, E. Heilman, M. L. Birch, & A. Mountain (Eds.), *Democratic responses in an era of standardization* (pp. 139–159). Troy, NY: Curriculum and Pedagogy.

Norris, J., Sawyer, R. D., & Lund, D. (Eds.). (2012). *Duoethnography: Dialogic methods for social, health, and educational research.* Walnut Creek, CA: Left Coast Press.

Norris, J., & Cope Watson, G. (2011, April). *Exploring the implicit dynamics of engendered authority: From elementary students to university professors.* Paper presented at the Annual Meeting of the American Educational Research Association, New Orleans, LA.

Oberg, A. (2002). Autobiographical research writing as a way of proceeding. Published as part of T. Wilson & A. Oberg, Side by side: Being in research autobiographically. *Educational Insights, 7*(2). Retrieved from http://

ccfi.educ.ubc.ca/publication/insights/v07n02/contextualexplorations/wilson_oberg/

Ondaatje, M. (1992). *The English patient*. New York, NY: Knopf.

The Oxford Online Dictionary. (2010). *Ethics*. Retrieved December 17, 2010, from http://oxforddictionaries.com/view/entry/m_en_us1245026#m_en_us1245026.

Pinar, W. F. (1975). Currere: Toward reconceptualization. In W. F. Pinar (Ed.), *Curriculum theorizing: The reconceptualists* (pp. 396–414). Berkeley, CA: McCutchan.

Pinar, W. F. (1978). Notes on the curriculum field 1978. *Educational Researcher*, *7*(8), 5–12.

Pinar, W. F. (1995). The method of currere (1975). In W. F. Pinar (Ed.), *Autobiography, politics and sexuality: Essays in curriculum theory 1972–1992* (pp. 19–27). New York, NY: Peter Lang.

Pinar, W. F. (2005). Curriculum. In S. J. Farenga, B. A. Joyce, & D. Ness (Eds.), *Encyclopedia on education and human development* (pp. 3–47). Armonk, NY: M. E. Sharpe.

Pinar, W. F., Reynolds, M. W., Slattery, P., & Taubman, M. P. (1995). *Understanding curriculum*. New York, NY: Peter Lang.

Poe, E. A. (1962). *Tales of mystery and imagination*. London, UK: Oxford University Press.

Polkinghorne, D. E. (1988). *Narrative knowing and the human sciences*. Albany, NY: State University of New York Press.

Prikryl, J. (2010, May 3). Extravagant disorder. *The Nation*, pp. 29–33.

Reason, P., & Hawkins, P. (1988). Storytelling as inquiry. In P. Reason (Ed.), *Human inquiry in action* (pp. 79–101). Newbury Park, CA: Sage.

Reason, P., & Rowan, J. (1981). Issues of validity in new paradigm research. In P. Reason & J. Rowan (Eds.), *Human inquiry: A sourcebook of new paradigm research* (pp. 239–262). New York, NY: John Wiley.

Reid, W. A. (1993). Does Schwab improve on Tyler? A response to Jackson. *Journal of Curriculum Studies*, *23*(6), 499–510.

Richardson, L. (1990). *Writing strategies: Reaching diverse audiences*. Newbury Park, CA: Sage.

Rosenblatt, L. (1978). *The reader, the text, the poem: The transactional theory of the literary work*. Carbondale, IL: Southern Illinois Press.

Roth, W. M. (Ed.). (2005). *Auto/biography and auto/ethnography: Praxis of research method*. Rotterdam, The Netherlands: Sense.

Routledge, P. (1996). The third space as critical engagement. *Antipode*, *28*(4), 399–419.

Rugg, H. O. (1936). *American life and the school curriculum: Next steps toward schools of living*. Boston, MA: Ginn.

Said, E. W. (1993). *Culture and imperialism*. New York, NY: Alfred A. Knopf.

Saukko, P. (2005). Methodologies for cultural studies: An integrative approach. In N. K. Denzin & Y. S. Lincoln (Eds.), *The SAGE handbook of qualitative research* (3rd ed., pp. 343–356). Thousand Oaks, CA: Sage.

Sawyer, R. D. (2010). Curriculum and international democracy: A Vital Source of Synergy and Change. *Journal of Curriculum Theorizing*, *26*(1), pp. 22–38.

Sawyer, R. D., & Laguardia, A. (2010). Reimagining the past/changing the present: Teachers adapting history curriculum for cultural encounters. *Teachers College Record, 112*(8), 1993–2020.

Sawyer, R. D., & Liggett, T. (2012). Postcolonial education: Using a duoethnographic lens to explore a personal curriculum of post/decolonization. In J. Norris, R. D. Sawyer, & D. Lund (Eds.), *Duoethnography: Dialogic methods for social, health, and educational research* (pp. 71–88). Walnut Creek, CA: Left Coast Press.

Sawyer, R. D., & Norris, J. (2009). Duoethnography: Articulations/(re)creation of meaning in the making. In W. R. Gershon (Ed.), *The collaborative turn: Working together in qualitative research* (pp. 127–140). Rotterdam, The Netherlands: Sense.

Scholes, R. (1985). *Textual power: Literary theory and the teaching of English.* New Haven, CT: Yale University Press.

Schwab, J. (1978). *Science, curriculum, and liberal education.* Chicago, IL: University of Chicago Press.

Schwandt, T. A. (2000). Three epistemological stances for qualitative inquiry: Interpretivism, hermeneutics, and social constructionism. In N. K. Denzin & Y. S. Lincoln (Eds.), *The SAGE handbook of qualitative research* (2nd ed., pp. 189–213). Thousand Oaks, CA: Sage.

Shelton, N. R., & McDermott, M. (2012). A curriculum of beauty. In J. Norris, R. D. Sawyer, & D. Lund (Eds.), *Duoethnography: Dialogic methods for social, health, and educational research* (pp. 223–242). Walnut Creek, CA: Left Coast Press.

Sitter, K., & Hall, S. (2012). Professional boundaries: Creating space and getting to the margins. In J. Norris, R. D. Sawyer, & D. Lund (Eds.), *Duoethnography: Dialogic methods for social, health, and educational research* (pp. 243–260). Walnut Creek, CA: Left Coast Press.

Smith, M. L., & Glass, G. V. (1987). *Research and evaluation in education and the social sciences.* Englewood Cliffs, NJ: Prentice-Hall.

Spradley, J. (1980). *Participant observation.* New York, NY: Holt, Rinehart and Winston.

Sullivan, G. (2005). *Art practice as research: Inquiry in the visual arts.* Thousand Oaks, CA: Sage.

Tolstoy, L. (1966). A talk among leisured people. In E. J. Simmons (Ed.), *Leo Tolstoy short novels* (Vol. 2, pp. 213–217). New York, NY: Modern Library.

Tyler, R. (1949). *Basic principles of curriculum and instruction.* Chicago, IL: University of Chicago Press.

van Manen, M. (1994). Pedagogy, virtue, and narrative identity in teaching. *Curriculum Inquiry, 24*(2), 135–170.

Varela, F. J., Thompson, E., & Rosch, E. (1991). *The embodied mind: Cognitive science and human experience.* Cambridge, MA: MIT Press.

Vygotsky, L. S. (1978). *Mind in society: The development of higher psychological processes.* Cambridge, MA: Harvard University Press.

Weber, M. (1949). *The methodology of the social sciences* (E. A. Shils & H. A. Finch, Eds. & Trans.). Chicago, IL: University of Chicago Press.

Wiebe, S., Sameshima, P., Irwin, R., Leggo, C., Gouzouasis, P., & Grauer, K. (2007). Rhizomatic relations of the everyday. *Journal of Educational Thought, 41*(3), 263–280.

Wiersma, W. (1991). *Research methods in education* (5th ed.). Needham Heights, MA: Allyn and Bacon.

Young, I. M. (1990). *Justice and the politics of difference.* Princeton, NJ: Princeton University Press.

General Readings

Autobiography/Autoethnography/Currere

Chang, H. (2008). *Autoethnography as method.* Walnut Creek, CA: Left Coast Press. This insightful book examines the diversity of autoethnography as a research method, placing it within historical, epistemological, and political perspectives and highlighting praxis and reflexivity currently emerging in this research genre.

Ellis, C. (2004). *The ethnographic I: A methodological novel about autoethnography.* Walnut Creek, CA: AltaMira Press. This book can be considered one of the foundational books that resulted in the acceptance of the use of the personal, the "I," in qualitative research.

Muncey, T. (2010). *Creating autoethnographies.* Thousand Oaks, CA: Sage. This book contains useful epistemological discussions and stylistic considerations for the writing of autoethnographies.

Pinar, W. F. (1975). Currere: Toward reconceptualization. In W. F. Pinar (Ed.), *Curriculum theorizing: The reconceptualists* (pp. 396–414). Berkeley, CA: McCutchan. This chapter articulated a call for the curriculum field to break away from an instrumental conflation of curriculum and instruction to begin to examine curriculum as living and meaningful text.

Pinar, W. F. (Ed.). (1994). *Autobiography, politics and sexuality: Essays in curriculum theory 1972–1992.* New York, NY: Peter Lang. This excellent book compiles Pinar's writings on curriculum theorizing. It contains an updated chapter on currere.

Curriculum

Aoki, T. T. (2005). *Curriculum in a new key: The collected works of Ted T. Aoki* (W. F. Pinar & R. L. Irwin, Eds.). Mahwah, NJ: Lawrence Erlbaum. In this exceptional collection of writings Ted Aoki presents stories of curricular landscapes of multiplicity, imagination found in a lived, relational curriculum.

Beyer, L. E., & Apple, M. W. (Eds.). *The curriculum: Problems, politics, and possibilities.* Albany, NY: State University of New York Press. This edited work presents a collection of essays that examine perceptions of curriculum within a range of foundational perspectives (e.g., historical, political, social, and theoretical).

Bradbeer, J. (1998). *Imagining curriculum: Practical intelligence in teaching.* New York, NY: Teachers College Press. This text explores the dimensions of lived curriculum, investigating how the concept of "mythopoesis" involves the experience and sources of cultural and personal negotiations of the imaginal.

Connelly, F. M., & Clandinin, D. J. (1988). *Teachers as curriculum planners: Narratives of experience.* New York, NY: Teachers College Press. This book examines the role of narrative and story in the idea and creation of curriculum, highlighting the meaning of a student-and-teacher lived curriculum.

Flinders, D. J., Noddings, N., & Thornton, S. J. (1986). The null curriculum: Its theoretical basis and practical implications. *Curriculum Inquiry, 16*(1), 33–42. This foundational paper makes explicit the power imbalances in the hidden curriculum (how we teach) and the null curriculum (what we don't teach).

Greene, M. (1978). *Landscapes of learning.* New York, NY: Teachers College Press. This book discusses the importance of personal and collective existential transactions for transcending taken-for-granted assumptions and patterned ways of seeing (and not seeing) the world.

Griffin, P., & Ouellett, M. L. (2007). Facilitating social justice education courses. In M. Adams, L. A. Bell, & P. Griffin (Eds.), *Teaching for diversity and social justice* (2nd ed., pp. 89–113). New York, NY: Routledge.

Grumet, M. R. (1988), *Bitter milk: Women and teaching.* Amherst, MA: University of Massachusetts Press. Drawing from feminist theory, Madeline Grumet's seminal text examines the role our bodies and embodied experience play in epistemological views of experience and curriculum.

Matus, C., & McCarthy, C. (2003). The triumph of multiplicity and the carnival of difference: Curriculum dilemmas in the age of postcolonialism and globalization. In W. F. Pinar (Ed.), *International handbook of curriculum research* (pp. 73–84). Mahwah, NJ: Lawrence Erlbaum. Matus and McCarthy examine emergent dilemmas found within the rapidly evolving and shifting landscapes of border-crossing multiplicity.

Pinar, W. F. (2005). Curriculum. In S. J. Farenga, B. A. Joyce, & D. Ness (Eds.), *Encyclopedia on education and human development* (pp. 3–47). Armonk, NY: M.E. Sharpe. In this comprehensive book chapter, William Pinar gives an exceptionally concise yet far-reaching presentation of the notion of the meaning and nature of curriculum (and curriculum theory), examining multiple, change-oriented conversations at its heart.

Duoethnography

Beidler, P. G., & Tong, R. (1994). Learning to teach. *Journal on Excellence in College Teaching, 5*(1), 107–126. While not referred to as a duoethnography, this early piece is written as a script of a conversation in which two people discuss their understanding of the act of teaching.

Norris, J., & Sawyer, R. (2004). Hidden and null curriculums of sexual orientation: A dialogue on the curreres of the absent presence and the present absence. In L. Coia, N. J. Brooks, S. J. Mayer, P. Pritchard, E. Heilman, M. L. Birch, & A. Mountain (Eds.), *Democratic responses in an era of standardization* (pp. 139–159). Troy, NY: Curriculum and Pedagogy. The first formal duoethnography, this study examines how two people are socially and culturally positioned in relation to their sexual identities.

Norris, J., Sawyer, R. D., & Lund, D. (Eds.). (2011). *Duoethnography: Dialogic methods for social, health, and educational research.* Walnut Creek, CA: Left Coast Press. This edited text presents 11 duoethnographic studies for social, health, and educational research. The editors provide analytical insight into the methodological value of each study.

Sawyer, R. D., & Norris, J. (2009). Duoethnography: Articulations/(re)creation of meaning in the making. In W. R. Gershon (Ed.), *The collaborative turn: Working together in qualitative research* (pp. 127–140). Rotterdam, The Netherlands: Sense. Using jazz as an organizing metaphor, this chapter examines the collaborative methodology of duoethnography as dialogic text.

Narrative Research

Barone, T. E. (1990). Using the narrative text as an occasion for conspiracy. In E. W. Eisner & A. Peshkin (Eds.), *Qualitative inquiry in education* (pp. 305–326). New York, NY: Teachers College Press. This chapter provides a strong argument for how stories enable readers to draw their own conclusions from evocative details rather than prescribed meanings of the authors.

Barone, T. (2001). Researching, writing and reading narrative studies. In T. Barone (Ed.), *Touching eternity: The enduring outcomes of teaching* (pp. 149–180). New York, NY: Teachers College Press. This chapter applies narrative research/studies to learning processes.

Barone, T. (2007). A return to the gold standard? Questioning the future of narrative construction as educational research. *Qualitative Inquiry, 13*(4), 454–470. This article discusses criteria by which narrative research can be assessed.

Clandinin, D. J., & Connelly, F. M. (2000). *Narrative inquiry: Experience and story in qualitative research.* San Francisco, CA: Jossey-Bass. This seminal book examines the epistemological and historical foundations of narrative inquiry and discusses ways of living and conducting it.

Philosophy/Epistemology/Ethics

Ayers, W., & Miller, J. L. (1997). *A light in dark times: Maxine Greene and the unfinished conversation.* New York, NY: Teachers College Press. This edited work presents a discussion of key writings of Maxine Greene by leading educational thinkers.

Bakhtin, M. M. (1981). *The dialogic imagination.* Austin, TX: University of Texas Press. This seminal text examines examples of dialogic imagination as found in literary texts. Key concepts include heteroglossia, dialogic change, and the carnival.

Bhabha, H. (1994). *The location of culture.* New York, NY: Routledge. This book explores the concept of the "third space"—the space created when two different cultures (or people) meet.

Dallery, A., & Scott, C. (1989). *The question of the Other.* Albany, NY: State University of New York Press. This collection of essays focuses on an ethical stance toward the Other. The "o" in Other is deliberately capitalized.

Fisher, G. (1996). *Gary in your pocket* (E. Sedgwick, Ed.). Durham, NC: Duke University Press. This is the edited diary of an African American man living in San Francisco in the 1990s, working on his PhD in english literature at the

University of California, Berkeley, and examining the meaning of life as a man living with HIV.

Kulp, C. B. (1992). *The end of epistemology*. This book discusses challenges that Dewey raised to the spectator theory of knowledge, including those found in science, instrumental thought, and epistemological puzzles.

Lévinas, E. (1984). Emmanuel Lévinas. In R. Kearney (Ed.), *Dialogues with contemporary Continental thinkers* (pp. 47–70). Manchester, UK: Manchester University Press. This chapter presents a conversation with Emmanuel Lévinas in which he summarizes his stance toward the world.

Prakash, M. S., & Esteva, G. (2008). *Escaping education: Living as learning in grassroots cultures*. New York, NY: Peter Lang. This book examines how epistemologies embedded in global instrumental notions of education are catastrophic to non-Western cultures and argues instead for alternative landscapes of learning within indigenous cultures.

Roth, W. M. (Ed.). (2005). *Auto/biography and auto/ethnography: Praxis of research method*. Rotterdam, The Netherlands: Sense. This book presents a variety of autoethnographies, which illustrate intersubjectivity, story, uncertainty, and praxis as research texts.

Said, E. (1993). *Culture and imperialism*. New York, NY: Alfred A. Knopf.

Qualitative Research

Bach, H. (1998). *A visual narrative concerning curriculum, girls, photography etc*. Edmonton, Alberta, Canada: International Institute for Qualitative Methodology. This work takes an arts-based approach to the exploration of teenage girls' narratives of experience.

Banks, A., & Banks, S. (1998). *Fiction and social research: By ice or fire*. Walnut Creek, CA: AltaMira Press. This book lays out an epistemological argument for how some fictive works can be considered qualitative research and then provides a number of chapter examples.

Creswell, J. (2007). *Qualitative inquiry and research design: Choosing among five approaches* (2nd ed.). Newbury Park, CA: Sage. This widely used qualitative research text includes narrative research as one of its research traditions.

Denzin, N., & Lincoln, Y. (Eds.). (2007). *Strategies of qualitative inquiry* (3rd ed.). Newbury Park, CA: Sage. This book contains a series of essays on various research orientations, including one called "Testimonio, Subalternity, and Narrative Authority." Earlier and later editions contain different chapters.

Gershon, W. R. (Ed.). (2009). *The collaborative turn: Working together in qualitative research*. Rotterdam, The Netherlands: Sense. This volume presents a diverse range of approaches to collaborative qualitative research, highlighting its meaning both to research and to researchers.

Leavy, P. (2009). *Method meets art*. New York, NY: Guilford Press. This insightful book discusses foundational contexts as well as methodological considerations of arts-based approaches to research.

Story

Haven, K. (2007). *Story proof: The science behind the startling power of story.* Westport, CT: Libraries Unlimited. This book summarizes a variety of research that validates the power of stories.

Reason, P., & Hawkins, P. (1988). Storytelling as inquiry. In P. Reason (Ed.), *Human inquiry in action* (pp. 79–101). Newbury Park, CA: Sage. This chapter is one of the seminal qualitative research pieces that lay the foundation for the value, purpose, and function of narrative research.

Organizations and Their Conferences

The following are organizations that hold conferences that either devote substantial focus to forms of narrative inquiry or are open to narrative texts, including duoethnographies.

International Congress of Qualitative Inquiry
http://www.icqi.org/
International Institute for Qualitative Methodology
http://www.iiqm.ualberta.ca/
Narrative Matters
http://w3.stu.ca/stu/sites/cirn/index.aspx
Centre for Qualitative Research
http://www.bournemouth.ac.uk/cqr/
Journal of Curriculum Theorizing—Bergamo Conference
http://www.jctonline.org/
Curriculum and Pedagogy
http://www.curriculumandpedagogy.org/Welcome.html
American Educational Research Association
http://www.aera.net/

Journals

The following journals have accepted narrative texts for publication.

International Journal of Education and the Arts
International Journal of Qualitative Methods
Qualitative Inquiry
Qualitative Research

INDEX